AF362031

Electronic Noses for Biomedical Applications and Environmental Monitoring

Electronic Noses for Biomedical Applications and Environmental Monitoring

Editor

Jesús Lozano

MDPI • Basel • Beijing • Wuhan • Barcelona • Belgrade • Manchester • Tokyo • Cluj • Tianjin

Editor
Jesús Lozano
Universidad de Extremadura
Spain

Editorial Office
MDPI
St. Alban-Anlage 66
4052 Basel, Switzerland

For citation purposes, cite each article independently as indicated on the article page online and as indicated below:

LastName, A.A.; LastName, B.B.; LastName, C.C. Article Title. *Journal Name* **Year**, *Volume Number*, Page Range.

ISBN 978-3-03943-937-9 (Hbk)
ISBN 978-3-03943-938-6 (PDF)

Contents

About the Editor

Jesús Lozano is a Full Professor at the University of Extremadura (SPAIN). He has been a research fellow at the Faculty of Physics of the Universidad Complutense de Madrid (UCM), research fellow at the Institute of Applied Physics of the Spanish Council for Scientific Research (CSIC), and lecturer in Naval Engineering of the Polytechnic University of Madrid (UPM). He received his degree in Industrial Engineering from the University of Extremadura (UEX) in 1998 and a degree and Ph.D. in Electronic Engineering. from the Universidad Complutense de Madrid (UCM) in 2001 and 2005, respectively. He also completed a Master of Science in Artificial Intelligence in 2009. Jesús Lozano is the author of more than 50 peer reviewed publications (h-index of 23, Scopus), 140 congress publications, 2 patents, 10 book chapters, and edited 8 books. He has also participated in more than 25 research projects related to gas sensors, electronic instrumentation, and intelligent systems.

He is now president of the Digital Olfaction Society and a member of the Spanish Olfaction Network, Iberian Network of Microsystems and Nanotechnology (Ibernam) and the Spanish Automation Committee (CEA).

Preface to "Electronic Noses for Biomedical Applications and Environmental Monitoring"

Electronic noses are bioinspired instruments that mimic the biological sense of smell. They are based on the use of gas sensors or biosensors combined with pattern recognition methods. Both topics have experienced great advances in the last few years: Chemical sensors have improved their metrological parameters, such as the limit of detection, the linearity of the response signal, sensitivity, selectivity, response time, and repeatability. The second involved the development of advanced embedded or remote signal and data analysis techniques, including big data and cloud computing. One of the main advantages of the use of electronic noses is the reduced cost, size, and ease of use, compared to traditional measurement systems, without the need for prior separation of the particular components of a gaseous mixture, which significantly reduces the time for a single analysis. For these reasons, the area of possible applications of electronic olfaction has been increasing over time. This Special Issue is devoted to the most recent technical developments in the area of electronic nose technology, including their design, chemical sensors and biosensors used, instrumentation systems for laboratory or field monitoring, personal systems and wearables, innovative data processing techniques, and also their implementation, in particular for biomedical applications and environmental monitoring.

Jesús Lozano
Editor

Article

Development of Tin Oxide-Based Nanosensors for Electronic Nose Environmental Applications

Isabel Sayago, Manuel Aleixandre and José Pedro Santos *

Institute of Physics Technology and Information (ITEFI-CSIC), 28006 Madrid, Spain; i.sayago@csic.es (I.S.);
manuel.aleixandre@csic.es (M.A.)
* Correspondence: jp.santos@csic.es; Tel.: +34-915-618-806

Received: 18 December 2018; Accepted: 28 January 2019; Published: 5 February 2019

Abstract: Tin oxide nanofibres (NFs) are used as nanosensors in electronic noses. Their performance is compared to that of oxide commercial chemical sensors for pollutant detection. NFs were grown by electrospinning and deposited onto silicon substrates with integrated micro-hotplates. NF morphology was characterized by scanning electron microscopy (SEM). The NFs presented high sensitivity to NO_2 at low temperature.

Keywords: nanofibres; tin oxide; electronic nose; NO_2; pollution; electrospinning; low detection temperature

1. Introduction

Pollution monitoring is the key to air quality management. The concentration of air pollutants is measured at network reference stations using precise analytical instruments consisting of bulky, heavy, and difficult-to-use high energy-consumption equipment. Thus, the number of available stations is limited due to high operation and maintenance costs. The stations are preferentially located in urban areas, although in many cases they are far from the main sources of pollution. However, in rural areas with few inhabitants and in remote or inaccessible areas, the measuring stations and therefore the pollution data are not available.

Currently, the most promising alternative for monitoring atmospheric pollutants is the use of electronic noses formed by a sensor array. The first step in the development of electronic noses for environmental applications is reducing the cost of sensors. These sensors are required, besides their low cost, to be autonomous, easy to use, reliable and accurate. Their size, weight, and energy consumption must also be reduced [1]. Resistive sensors of metal oxide semiconductors (MOX) are suitable candidates for the development of low-cost, high-performance sensors due to the simplicity of the physical magnitude involved in the measurement (resistance) and the high sensibility to toxic gases. In particular, nanostructured materials are the most appropriate strategy to minimize some of the current problems with gas sensors (lack of sensitivity, power consumption, and stability).

In this work, we present the development of tin oxide nanosensors for electronic noses (e-noses). The two main applications of e-noses in the environment are pollution and odour monitoring. Due to the increased interest in this field and in order to improve potential use of instrumental odour monitoring, including sensors or e-noses, a new working group (WG41) started in 2015 within the framework of the European Committee of Standardization (CEN/TC264 Air Quality). The objective of this group was to propose a new European standard for instrumental odour monitoring [2]. Applications of electronic noses in the environment can be found in several works, some based on MOX [3] or amperimetric commercial sensors [4]. Other types of e-noses are those based on surface acoustic wave (SAW) sensors [5]. Biomimetic artificial noses, including whole-cell olfactory receptor protein and odorant binding protein (OBP)-based biosensors are also being studied [6]. Portable devices are being developed for the measurement of urban pollution [7–9].

Gas sensors based on sensitive layers of one-dimensional metal oxide (1D) nanostructures have shown superior performance to bulk sensors due to their large surface area–volume ratio and their dimensions being comparable to the extent of the surface charge region [10–12]. Tin oxide is still the most important material used for the detection of atmospheric polluting gases, and its most outstanding characteristics with respect to other semiconductors are its high sensitivity at low temperatures and low cost. One-dimensional SnO_2 nanostructures (nanowires, nanobelts, nanoribbons, nanofibres, etc) can be synthesized using several methods like laser ablation, chemical vapour deposition, electro-deposition, thermal evaporation, rapid oxidation and electrospinning [13–16].

Electrospinning is a simple, versatile and economic technique that allows fibres to be obtained at micro and nanometric scales [17,18]. The electrospinning process began to be employed in conventional organic polymers of high molecular weight [19] and in the last decade has been used for the preparation of semiconductor oxide fibres from polymer solutions incorporating metallic precursors [20–22]. The process involves the application of an electrostatic field to a polymer solution with a certain viscosity and when the electric field strength is greater than the surface tension, the polymer solution is expelled to a collector in the form of a fibre.

Nitrogen dioxide (NO_2) is one of the major air pollutants, especially in large cities. NO_2 is an oxidizing gas whose main emission sources are combustion processes (heating, power generation and engines in vehicles and ships). Its effect on human health can be both short-term (causing significant inflammation of the respiratory tract) and long-term (affecting organs such as the liver and spleen, systems such as the circulatory system and the immune system, which in turn leads to lung infections and respiratory failure) [23]. In addition, nitrogen oxides alter the environment by contributing to the acidification and eutrophication (excess nitrogen nutrients) of terrestrial and aquatic ecosystems, leading to a loss of life in animals and plants and changes in species diversity [24].

The NO_2 exposure limit values recommended by the World Health Organization (WHO) [25] are shown in Table 1. These low concentration ranges cannot be detected by commercial sensors at low temperature.

Table 1. NO_2 limit values recommended by the WHO.

Average Annual	Average Hourly
40 μg/m^3 (0.02 ppm)	200 μg/m^3 (0.11 ppm) not to exceed more than 18 h per year

The European Commission [26] has urged member states to implement air quality management plans that ensure compliance with the standards set by the EU air quality directive [27] no later than 2020. Air pollution monitoring is a key air quality management task, for which the Air Quality Directive (AQD) opts for a strategy based on a network of a limited number of fixed stations, equipped with precision analytical instruments, which has some drawbacks.

Measuring equipment is bulky, heavy, difficult to use, and consumes a lot of energy. Equipment costs, operation, and maintenance are high. In many cases, the stations are located far away from areas of high traffic density where the greatest local increases in air pollution occur. A small number of these stations dispersed in a city allows data to be obtained with hourly resolution, but at a small number of points. In emergency situations, decisions are based on real-time measurements or, in the absence of such measurements, on predictive models of the distribution of pollutants in the atmosphere, the usefulness of which depends on the degree of validation of the models. Thus, although stations accurately measure air pollution, their spatial representativeness and temporal resolution are insufficient to capture the spatial–temporal variability of air pollution.

Although the AQD does not consider sensors as reference instruments, it does open the door to the use of sensors for indicative measurements, for which it sets less restrictive quality objectives. It is estimated that the use of low-cost, low-consumption sensors that meet AQD quality standards for indicative measurements would allow a 50% reduction in the minimum number of stations [28].

The new generation of sensors finds application (unregulated) in sectors such as personal and community monitoring of air quality, traffic management, estimation of exposure to air pollution, R&D, and environmental education, in which there are numerous business opportunities.

In this work, two prototypes of electronic noses for environmental applications based on low-cost sensors are described. The low-cost sensors tested were nanostructured tin oxide materials (nanofibres) obtained by an economical and versatile process (electrospinning) and commercial sensors. The sensor responses to low concentrations of NO_2 [29] in controlled air atmospheres are also presented and discussed. We obtained good responses even at room temperature.

These nanofibre-based tin oxide resistive sensors can be incorporated into an electronic nose and could be used for air quality control.

2. Materials and Methods

2.1. Materials

Polyvinyl alcohol (PVA) and tin chloride (II) pentahydrate ($SnCl_4 \cdot 5H_2O$) were used as precursor materials and distilled water was used as a solvent. PVA with an average molecular weight of 80,000 g/mol and $SnCl_4 \cdot 5H_2O$ were supplied by Sigma–Aldrich Química (Madrid, Spain).

2.2. Preparation of Precursor Solution

First, an aqueous PVA solution (11% wt.) was prepared by dissolving PVA in distilled water and heating at 80 °C, under stirring for 2 h. Next, $SnCl_4 \cdot 5H_2O$ was added and the solution was cooled to room temperature, with stirring during cooling.

2.3. Synthesis of Tin Oxide Nanofibres

The SnO_2 nanofibres (NFs) were prepared by an electrospinning process. The precursor solution (PVA + $SnCl_4 \cdot 5H_2O$) was loaded into a syringe equipped with a metallic needle. A positive voltage of 19 kV was applied to the needle tip and the metal collector was grounded. The solution flow rate was 2 µL/min and the distance between the needle tip and the collector (silicon substrate) was 6 cm. Details of the electrospinning system are described in a previous work [30]. All sensors were prepared in the same conditions and NFs were grown onto micromachined silicon substrates with integrated heaters that allowed the calcination of the nanofibres in the test cell. The NFs were calcined at 500 °C for 4 h in air, obtaining nanofibres of SnO_2.

2.4. Experimental Setup of the E-Nose System

Electronic Noses

Two electronic noses were developed: WiNOSE 5.0 for the nanosensors (R1, R2 and R3 nanofibre-based tin oxide sensors) and WiNOSE 6.0 for the commercial sensors. The schematics of both e-noses were very similar. The main difference between them is that the former is intended for laboratory use and the latter is a hand-held device that can also be used in the field [31]. Figure 1 shows the schematics of the WiNOSE. Details of the electronics can be found in [32].

The gases were generated by the dynamic dilution of bottles of 2 ppmv of NO_2 in synthetic air. The sensors were heated to several temperatures using the resistances integrated into the micro-machined sensors and controlled by the electronic nose. The electronic nose and the gas generation instrumentation were controlled by a custom LabVIEW software that also registered the measurements to a computer. Figure 2 shows the scheme of the experimental set-up to measure the sensors.

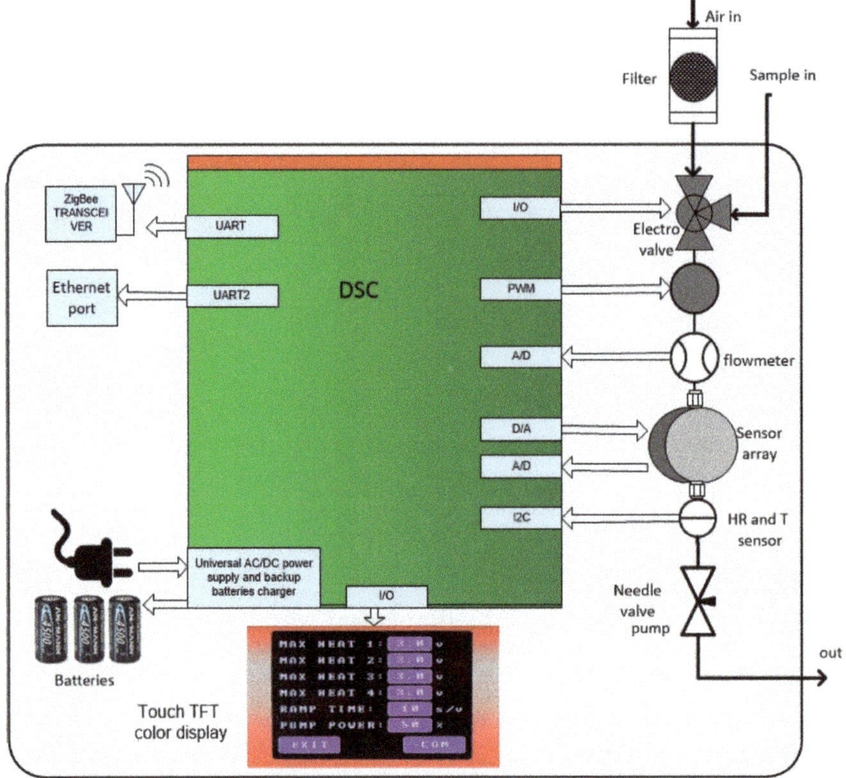

Figure 1. WiNOSE schematics.

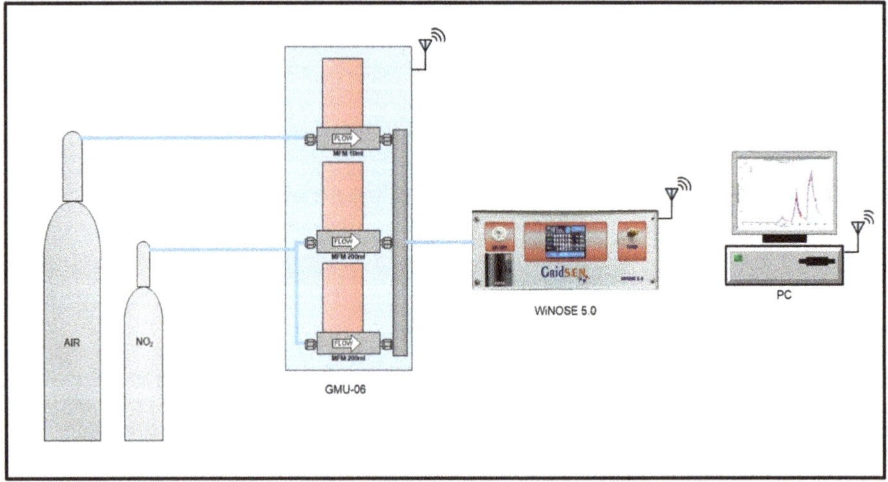

Figure 2. Scheme of the experimental design to measure the sensors.

Detections were carried out in air at temperatures ranging from 25 to 200 °C, with a constant flow of 200 mL/min. The NO_2 concentrations varied from 0.1 to 0.5 ppmv with an exposure time of 10 min.

2.5. Sensor Tested

The WiNOSE 5.0 using three tin oxide NF nanosensors (R1, R2 and R3) was prepared with the same procedure. The silicon substrates of the sensors had integrated microheaters that allowed the sensitive layers to be heated and interdigitated electrodes (IDTs) to measure the sensitive layer resistance. The substrates with sensitive layers of tin oxide NFs covering the surface of the IDTs were mounted in a standard TO-8 package for the electrical characterization of the sensors. The TO-8 device was placed in the stainless-steel test cell inside the apparatus.

The WiNOSE 6.0 uses eight state-of-the-art commercial metal-oxide (MOX) microsensors, CC801 and CC803 (Cambridge CMOS Sensors Ltd., Cambridge, UK), operating at different temperatures. CC801 is intended for monitoring indoor air quality including carbon monoxide (CO) and a wide range of volatile organic compounds (VOCs), while CC803 is aimed at the detection of ethanol. However, like the majority of MOX sensors, they are also sensitive to NO_2.

3. Results

3.1. Morphological Characterization of Tin Oxide Nanofibres

The fibres were randomly distributed on the substrate forming porous interlaced networks, as can be seen in the SEM images (Figure 3). In general, the fibres had nanometric diameters from 40 to 50 nm and their lengths reached several tens of microns. The nanofibres were constituted by multitude of nanograins whose diameters were less than 15 nm, as calculated from the broadening of the X-ray diffraction peaks in a previous work [30]. The nanograins were evenly distributed in the fibres, forming a porous nanostructure (Figure 3b).

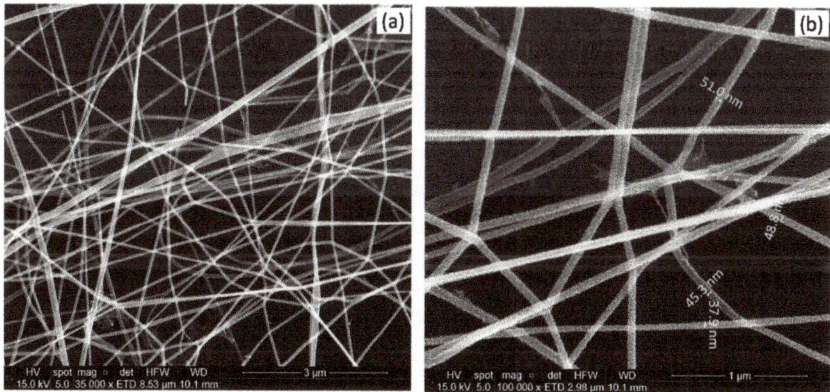

Figure 3. SEM micrographs of tin oxide NFs produced by electrospinning after calcination (500 °C in air for 4 h). (**a**) magnification 35000; (**b**) magnification 100000.

3.2. WiNOSE 5.0

The WiNOSE 5.0 consists of the three tin oxide NF nanosensors (R1, R2 and R3). The sensors were exposed to different NO_2 concentrations in the sub-ppmv range (0.1 to 1 ppmv). Figure 4 shows the sensor resistance changes in the detection processes. At room temperature, the resistance changed only with concentrations higher than 0.1 ppmv NO_2. However, at 150 and 200 °C, the sensor detected 0.1 ppmv NO_2 with a response ($R = (R/R_a)$, where R_a and R stand for the sensor resistance in air and under exposure to NO_2, respectively) of 1.42 and 1.37, respectively. While the responses were high at temperatures below 200 °C, the response times were slow. At low temperature, the sensors did not reach saturation during the exposure time to NO_2, although the resistance changes were observed after 2 min of exposure. Both the response and recovery processes depended on the operating temperature. At 200 °C, the responses obtained were lower than those reached at 150 °C. However, the sensors reached saturation at 200 °C during exposure to NO_2 and at this temperature, the response and recovery times were lower than at 150 °C.

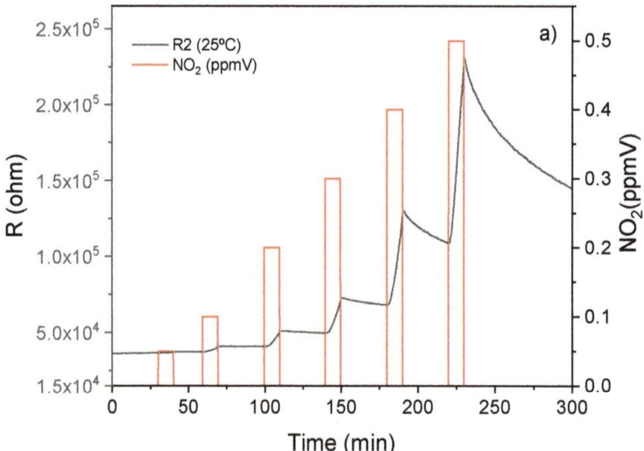

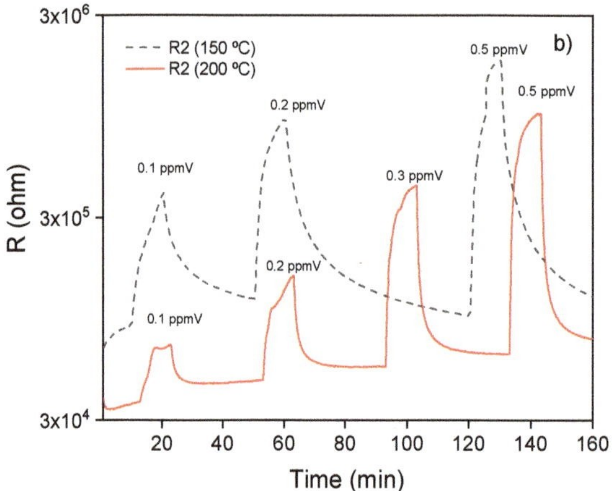

Figure 4. Response curves of the R2 sensor to NO_2 at different temperatures: (**a**) room temperature and (**b**) 150 and 200 °C.

Figure 5 shows the responses achieved in the detection of 0.1, 0.2, 0.5, and 1 ppmv NO_2 at different temperatures with the R1 and R2 sensors. No remarkable differences were observed. The response of the R3 sensor was very similar to that of the other two. All sensors tested had a maximum sensitivity at 150 °C. Therefore, the optimum detection temperature may be between 150 and 200 °C.

In order to check the long-term repeatability and reliability of the sensors, the detections were repeated after 10 weeks. The response curves obtained after inactive periods were similar (Figure 6), which confirms the reproducibility of the results. At 10 weeks, an increase of the sensors' resistance was observed due to a slow aging process via interaction with surrounding gases. These increases were more evident as the operating temperature of the sensor became higher.

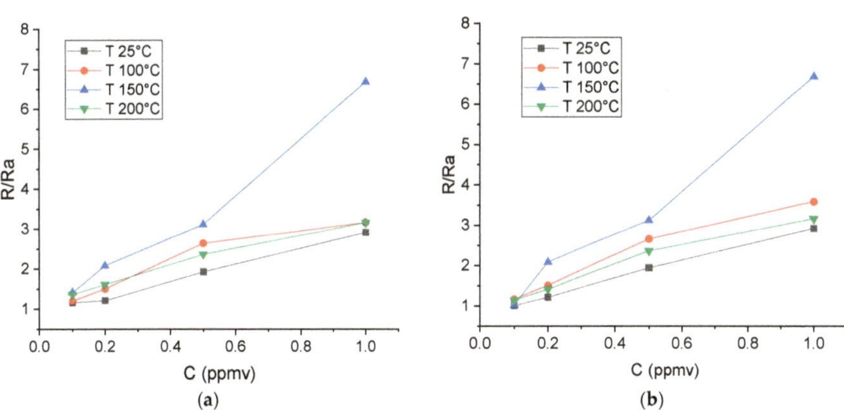

Figure 5. Response of the R1 (**a**) and R2 (**b**) sensors (sensitive layer of tin oxide NFs) to low NO_2 concentrations at different operating temperatures.

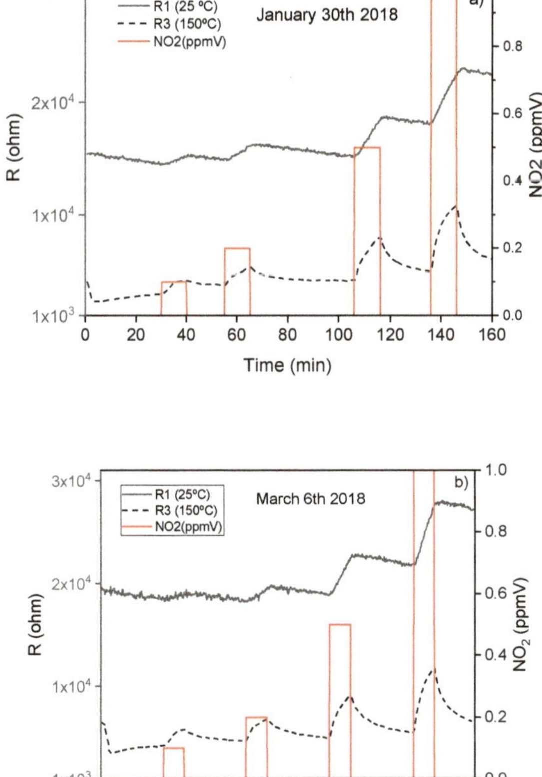

Figure 6. Repeatability of the R1 and R3 sensors exposed to different concentrations of NO_2. Response curves of R1and R3: (**a**) initially, (**b**) after 10 weeks.

Calibrations of Nanosensors

The responses of the R1, R2 and R3 sensors to NO_2 were measured at 25 °C, 50 °C, 100 °C and 200 °C to determine the best operating temperature and the possibility of operating these sensors at low temperatures. In order to study more accurately the performance of the sensors, calibration curves were calculated and analysed for various temperatures. The responses of two of the sensors (R1 and R3) tested for NO_2 detection (between 0.1 and 2 ppmv) were used for calibration. An ortho-normal calibration [33] was performed and the RMS and R^2 of the calibrations were calculated, as shown in Figure 7. This calculation was repeated for each sensor and each temperature, and the results are compiled in Table 2. To test the combined power of the two sensors, we also carried out a partial least squares (PLS) regression with both sensors as independent variables and the concentration of NO_2 as the dependent variable. The PLS was validated and evaluated by leave-one-out cross validation. This validation consisted of a loop in which every point was selected once. Then the rest of the points were used to compute a calibration that was used to predict the concentration of the point left out. This prediction was compared with the real concentration. The results can also be seen in Table 2.

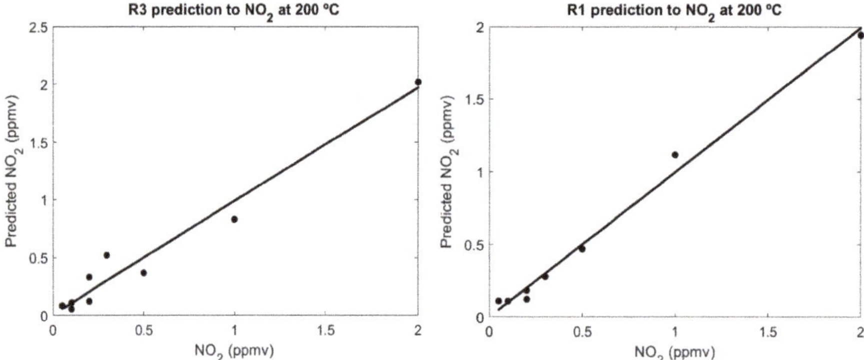

Figure 7. Calibration of the R1 and R3 sensor at 200 °C.

Table 2. Errors of the different calibrations.

T (°C)	RMS R1	RMS R3	RMS PLS	R^2 R1	R^2 R3	R^2 PLS
25	0.186	0.0288	0.328	0.993	0.997	0.992
50	0.159	0.090	0.140	0.989	0.989	0.998
100	0.106	0.059	0.267	0.998	0.987	0.979
150	0.539	0.025	0.473	0.999	0.995	0.969
200	0.119	0.246	0.772	0.954	0.996	0.840
250	0.321	0.034	0.176	0.903	0.999	0.975
300	0.345	0.227	2.212	0.976	0.964	0.864

3.3. WiNOSE 6.0

Measurements of low NO_2 concentrations were performed at several temperatures ranging from 20 °C to 350 °C. Meaningful responses were only obtained above 250 °C. Figure 8 shows the response of the two types of commercial sensors at 255 °C and 350 °C. The same analysis as for the calibration was carried out for the commercial sensors and the results are summarized in Table 3.

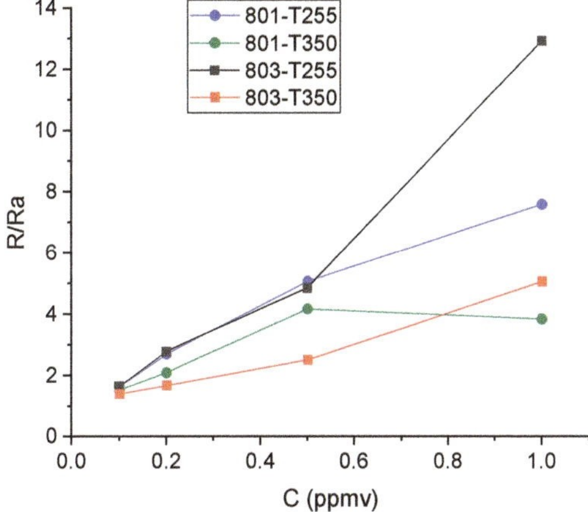

Figure 8. Response of the commercial sensors to low NO_2 concentrations at two operating temperatures.

Table 3. Errors of the different calibrations for the commercial sensors.

T (°C)	RMS S801	RMS S803	RMS PLS	R^2 S801	R^2 S803	R^2 PLS
255	0.008	0.043	1.197	0.990	0.987	0.941
350	0.372	0.006	0.695	0.949	0.989	0.984

4. Discussion

In the detection processes, the resistance changes occurred with the adsorption of gaseous molecules on the sensitive surface. Nanostructures were considered for gas detection applications due to their high surface area–volume ratio. In this work, the nanostructures—porous nanofibre networks—were composed of many nanograins that favoured the adsorption of gases.

The sensor calibrations had low errors, especially around 50–100 °C, and a lower error at room temperature. At higher temperatures, the sensors probably experienced some instability and the measurements had a much higher variability, which reflected the weaker performance. The sensors showed a good linear response in the concentration range tested. The combination of both sensors in a multilinear calibration was validated and the results were better estimated because the stricter validation and the aggregation of both sensors on a single performance was validated. The PLS had low error that tended to increase with the temperature and showed very good performance at 50 °C.

The sensors based on nanofibres had better low-temperature performance than commercial sensors and also better than that reported in the literature (Table 4). The references showed that NO_2 concentrations lower than 0.5 ppm were detected and that the sensors would operate at moderate temperatures generally higher than 150 °C. Most of the references of the sensitive layers corresponded to complex nanostructures prepared by hydrothermal methods (due to difficulty to control the process, and problems of reliability and reproducibility). Although there are usually references for the sensor response (R_{NO2}/R_{air}), there is no detail of the sensor resistance. The commercial sensors used in this work, did not have any significant response below 250 °C, but they showed a more stable response with lower errors in the calibration for higher temperatures.

Table 4. Comparison of NO_2-resistive gas sensors based in nanostructured MOX.

Sensitive Layer	Concentration (ppm)	T ($^\circ$C)	Response (R_{NO2}/R_{air})	Ref.
In_2O_3 (nanorod clusters)	0.5	150	41	[34]
ZnO (nanowires)	0.5	225	18	[35]
SnO_2 (nanowires)	0.5	200	17	[36]
SnO_2 (hierarchical leaf-like)	0.5	65	7	[37]
SnO_2 (nanofibrefibres)	0.1/0.5	25	1.16/1.93	This work

5. Conclusions

The results confirm that electrospun tin oxide nanostructured sensors can be used as sensors in electronic noses for environmental applications due to their high response to low NO_2 concentrations, even at room temperature. They will allow for the development of new low-cost, low-consumption, sensor-based smart systems for the detection of gases. The adequate distribution of sensor networks (electronic noses) can provide information on pollution variation in large areas.

In future work, the humidity effect and ozone interference on sensor responses will be studied. In order to improve the sensor performance, catalytic metals (Au, Pd, and Ag) or graphene will be incorporated into the nanofibres. These additives will increase the sensor response at low temperatures and accelerate the processes of absorption and desorption.

Author Contributions: Methodology, I.S. and J.P.S.; Software, M.A.; Formal Analysis, M.A.; Investigation, I.S., J.P.S. and M.A.; Data Curation, M.A.; Writing—Original Draft Preparation, I.S.; Writing—Review & Editing, I.S., J.P.S. and M.A.

Acknowledgments: SEM analyses were performed by the University of Extremadura.

Conflicts of Interest: The authors declare no conflict of interest.

References

1. Low-Cost Sensors for the Measurement of Atmospheric Composition: Overview of Topic and Future Applications | Climate & Clean Air Coalition. Available online: http://www.ccacoalition.org/en/resources/low-cost-sensors-measurement-atmospheric-composition-overview-topic-and-future (accessed on 25 October 2018).

2. Guillot, J.-M. E-Noses: Actual Limitations and Perspectives for Environmental Odour Analysis. *Chem. Eng. Trans.* **2016**, *54*, 223–228.

3. Szulczyński, B.; Wasilewski, T.; Wojnowski, W.; Majchrzak, T.; Dymerski, T.; Namieśnik, J.; Gębicki, J. Different Ways to Apply a Measurement Instrument of E-Nose Type to Evaluate Ambient Air Quality with Respect to Odour Nuisance in a Vicinity of Municipal Processing Plants. *Sensors* **2017**, *17*, 2671. [CrossRef] [PubMed]

4. Jasinski, G.; Wozniak, L.; Kalinowski, P.; Jasinski, P. Evaluation of the Electronic Nose Used for Monitoring Environmental Pollution. In Proceedings of the 2018 IEEE XV International Scientific Conference on Optoelectronic and Electronic Sensors (COE), Orlando, FL, USA, 10–13 November 2018; pp. 1–4.

5. Panneerselvam, G.; Thirumal, V.; Pandya, H.M. Review of Surface Acoustic Wave Sensors for the Detection and Identification of Toxic Environmental Gases/vapours. *Arch. Acoust.* **2018**, *43*, 357–367.

6. Dung, T.; Oh, Y.; Choi, S.-J.; Kim, I.-D.; Oh, M.-K.; Kim, M.; Dung, T.T.; Oh, Y.; Choi, S.-J.; Kim, I.-D.; et al. Applications and Advances in Bioelectronic Noses for Odour Sensing. *Sensors* **2018**, *18*, 103. [CrossRef] [PubMed]

7. Capezzuto, L.; Abbamonte, L.; De Vito, S.; Massera, E.; Formisano, F.; Fattoruso, G.; Di Francia, G.; Buonanno, A. A Maker Friendly Mobile and Social Sensing Approach to Urban Air Quality Monitoring. In Proceedings of the 2014 IEEE Sensors, Valencia, Spain, 2–5 November 2014.

8. Hannon, A.; Lu, Y.; Li, J.; Meyyappan, M. A Sensor Array for the Detection and Discrimination of Methane and Other Environmental Pollutant Gases. *Sensors* **2016**, *16*, 1163. [CrossRef] [PubMed]

9. Herrero, J.L.; Lozano, J.; Santos, J.P.; Fernandez, J.A.; Marcelo, J.I.S. A Web-Based Approach for Classifying Environmental Pollutants Using Portable E-Nose Devices. *IEEE Intell. Syst.* **2016**, *31*, 108–112. [CrossRef]

10. Sun, Y.-F.; Liu, S.-B.; Meng, F.-L.; Liu, J.-Y.; Jin, Z.; Kong, L.-T.; Liu, J.-H.; Sun, Y.-F.; Liu, S.-B.; Meng, F.-L.; et al. Metal Oxide Nanostructures and their Gas Sensing Properties: A Review. *Sensors* **2012**, *12*, 2610–2631. [CrossRef] [PubMed]

11. Comini, E.; Baratto, C.; Faglia, G.; Ferroni, M.; Vomiero, A.; Sberveglieri, G. Quasi-One Dimensional Metal Oxide Semiconductors: Preparation, Characterization and Application as Chemical Sensors. *Prog. Mater. Sci.* **2009**, *54*, 1–67. [CrossRef]

12. Pan, J.; Shen, H.; Mathur, S. One-Dimensional SnO$_2$ Nanostructures: Synthesis and Applications. *J. Nanotechnol.* **2012**, *2012*, 1–12. [CrossRef]

13. Yu, F.; Tang, D.; Hai, K.; Luo, Z.; Chen, Y.; He, X.; Peng, Y.; Yuan, H.; Zhao, D.; Yang, Y. Fabrication of SnO$_2$ One-Dimensional Nanosturctures with Graded Diameters by Chemical Vapor Deposition Method. *J. Cryst. Growth* **2010**, *312*, 220–225. [CrossRef]

14. Luan, C.; Zhu, Z.; Mi, W.; Ma, J. Structural and Electrical Properties of SnO$_2$ Films Grown on R-Cut Sapphire at Different Substrate Temperature by MOCVD. *Vacuum* **2014**, *99*, 110–114. [CrossRef]

15. Kim, J.-H.; Abideen, Z.U.; Zheng, Y.; Kim, S.S. Improvement of Toluene-Sensing Performance of SnO$_2$ Nanofibers by Pt Functionalization. *Sensors* **2016**, *16*, 1857. [CrossRef] [PubMed]

16. Park, J.Y.; Asokan, K.; Choi, S.-W.; Kim, S.S. Growth Kinetics of Nanograins in SnO$_2$ Fibers and Size Dependent Sensing Properties. *Sens. Actuators B Chem.* **2011**, *152*, 254–260. [CrossRef]

17. Kenry; Lim, C.T. Nanofiber Technology: Current Status and Emerging Developments. *Prog. Polym. Sci.* **2017**, *70*, 1–17. [CrossRef]

18. Li, D.; McCann, J.T.; Xia, Y.; Marquez, M. Electrospinning: A Simple and Versatile Technique for Producing Ceramic Nanofibers and Nanotubes. *J. Am. Ceram. Soc.* **2006**, *89*, 1861–1869. [CrossRef]

19. Huang, Z.-M.; Zhang, Y.-Z.; Kotaki, M.; Ramakrishna, S. A Review on Polymer Nanofibers by Electrospinning and Their Applications in Nanocomposites. *Compos. Sci. Technol.* **2003**, *63*, 2223–2253. [CrossRef]

20. Lim, S.K.; Hwang, S.-H.; Chang, D.; Kim, S. Preparation of Mesoporous In$_2$O$_3$ Nanofibers by Electrospinning and Their Application as a CO Gas Sensor. *Sens. Actuators B Chem.* **2010**, *149*, 28–33. [CrossRef]

21. Kim, I.-D.; Rothschild, A. Nanostructured Metal Oxide Gas Sensors Prepared by Electrospinning. *Polym. Adv. Technol.* **2011**, *22*, 318–325. [CrossRef]

22. Ding, B.; Wang, M.; Yu, J.; Sun, G.; Ding, B.; Wang, M.; Yu, J.; Sun, G. Gas Sensors Based on Electrospun Nanofibers. *Sensors* **2009**, *9*, 1609–1624. [CrossRef] [PubMed]

23. Latza, U.; Gerdes, S.; Baur, X. Effects of nitrogen dioxide on human health: Systematic review of experimental and epidemiological studies conducted between 2002 and 2006. *Int. J. Hyg. Environ. Health* **2009**, *212*, 271–287. [CrossRef] [PubMed]

24. Air Quality in Europe—2018 Report. Available online: https://www.eea.europa.eu/publications/air-quality-in-europe-2018 (accessed on 15 november 2018).

25. Ambient (Outdoor) Air Quality and Health. Available online: http://www.who.int/news-room/fact-sheets/detail/ambient-(outdoor)-air-quality-and-health (accessed on 25 October 2018).

26. European Environment Agency. Communication from the Commission to the Council, the European Parliament, the European Economic and Social Committee and the Committee of the Regions—"A Clean Air Programme for Europe", COM (2013) 918 final. Available online: http://eur-lex.europa.eu/legal-content/EN/TXT/PDF/?uri=CELEX:52013DC0918&from=EN (accessed on 1 February 2019).

27. European Environment Agency. Directive 2008/50/EC of the European Parliament and of the Council of 21 May 2008 on ambient air quality and cleaner air for Europe. Available online: http://eur-lex.europa.eu/legal-content/en/ALL/?uri=CELEX:32008L0050 (accessed on 1 February 2019).

28. TD1105—European Network on New Sensing Technologies for Air-Pollution Control and Environmental Sustainability—EuNetAir. Available online: https://www.cost.eu/actions/TD1105/#tabs\T1\textbar{}Name:overview (accessed on 1 February 2019).

29. Spinelle, L.; Gerboles, M.; Villani, M.G.; Aleixandre, M.; Bonavitacola, F. Field Calibration of a Cluster of Low-Cost Available Sensors for Air Quality Monitoring. Part A: Ozone and Nitrogen Dioxide. *Sens. Actuators B Chem.* **2015**, *215*, 249–257. [CrossRef]

30. Santos, J.; Fernández, M.; Fontecha, J.; Matatagui, D.; Sayago, I.; Horrillo, M.; Gracia, I.; Santos, J.P.; Fernández, M.J.; Fontecha, J.L.; et al. Nanocrystalline Tin Oxide Nanofibers Deposited by a Novel Focused Electrospinning Method. Application to the Detection of TATP Precursors. *Sensors* **2014**, *14*, 24231–24243. [CrossRef] [PubMed]

31. Santos, J.P.; Aleixandre, M.; Arroyo, P.; Suárez, J.I.; Lozano, J. An Advanced Hand Held Electronic Nose for Ambient Air Applications. *Chem. Eng. Trans.* **2018**, *68*, 235–240.

32. Santos, J.P.; Aleixandre, M.; Horrillo, M.C. WiNOSE: Wireless Electronic Nose for Outdoors Applications. *Chem. Eng. Trans.* **2010**, *23*, 159–164.

33. Han, D.; Zhai, L.; Gu, F.; Wang, Z. Highly sensitive NO_2 gas sensor of ppb-level detection based on In_2O_3 nanobricks at low temperature. *Sens. Actuators B Chem.* **2018**, *262*, 655–663. [CrossRef]

34. Xu, X.; Zhang, H.; He, C.; Pu, C.; Leng, Y.; Li, G.; Hou, S.; Zhu, Y.; Fu, L.; Lu, G. Synthesis and NO_2 Sensing Properties of Indium Oxide Nanorod Clusters via a Simple Solvothermal Route. *RSC Adv.* **2016**, *6*, 47083–47088. [CrossRef]

35. Ahn, M.-W.; Park, K.-S.; Heo, J.-H.; Kim, D.-W.; Choi, K.J.; Park, J.-G. On-Chip Fabrication of ZnO-Nanowire Gas Sensor with High Gas Sensitivity. *Sens. Actuators B Chem.* **2009**, *138*, 168–173. [CrossRef]

36. Choi, Y.-J.; Hwang, I.-S.; Park, J.-G.; Choi, K.J.; Park, J.-H.; Lee, J.-H. Novel Fabrication of an SnO_2 Nanowire Gas Sensor with High Sensitivity. *Nanotechnology* **2008**, *19*, 95508. [CrossRef] [PubMed]

37. Zhang, Y.; Li, D.; Qin, L.; Zhao, P.; Liu, F.; Chuai, X.; Sun, P.; Liang, X.; Gao, Y.; Sun, Y.; et al. Preparation and Gas Sensing Properties of Hierarchical Leaf-like SnO_2 Materials. *Sens. Actuators B Chem.* **2018**, *255*, 2944–2951. [CrossRef]

Article

Innovative Sensor Approach to Follow *Campylobacter jejuni* Development

Estefanía Núñez-Carmona [1], Marco Abbatangelo [1,*] and Veronica Sberveglieri [2,3]

[1] Department of Information Engineering, University of Brescia, Brescia, via Branze, 38, 25123 Brescia, BS, Italy; e.nunezcarmona@unibs.it

[2] CNR-IBBR, Institute of Bioscience and Bioresources, via Madonna del Piano, 10, 50019 Sesto Fiorentino, FI, Italy; veronica.sberveglieri@ibbr.cnr.it

[3] NANO SENSOR SYSTEMS, NASYS Spin-Off University of Brescia, Brescia, via Camillo Brozzoni, 9, 25125 Brescia, BS, Italy

* Correspondence: m.abbatangelo@unibs.it

Received: 7 November 2018; Accepted: 26 December 2018; Published: 7 January 2019

Abstract: *Campylobacter spp* infection affects more than 200,000 people every year in Europe and in the last four years a trend shows an increase in campylobacteriosis. The main vehicle for transmission of the bacterium is contaminated food like meat, milk, fruit and vegetables. In this study, the aim was to find characteristic volatile organic compounds (VOCs) of *C. jejuni* in order to detect its presence with an array of metal oxide (MOX) gas sensors. Using a starting concentration of 10^3 CFU/mL, VOCs were analyzed using Gas-Chromatography Mass-Spectrometry (GC-MS) with a Solid-Phase Micro Extraction (SPME) technique at the initial time (T0) and after 20 h (T20). It has been found that a Campylobacter sample at T20 is characterized by a higher number of alcohol compounds that the one at T0 and this is due to sugar fermentation. Sensor results showed the ability of the system to follow bacteria curve growth from T0 to T20 using Principal Component Analysis (PCA). In particular, this results in a decrease of $\Delta R/R_0$ value over time. For this reason, MOX sensors are a promising technology for the development of a rapid and sensitive system for *C. jejuni*.

Keywords: *Campylobacter jejuni*; VOCs; GC-MS SPME; nanowire sensors; PCA

---

1. Introduction

Nowadays we live in the safest environment regarding the food industry. Organizations as EFSA (European Food Safety Agency), WHO (World Health Organization), and FAO (Food and Agriculture Organization) determine, organize and rule all the regulations that control every single aspect of food safety, security and trading. Even if the risk perception regarding the food industry is not one of the first concerns of the population in non-developing countries [1,2], food poisoning is still the first cause of hospitalization in the world. The CDC (Center of Disease Control) estimates that each year 48 million people get sick from a foodborne illness, 128,000 are hospitalized and 3000 die just in the USA [3].

The FoodNet (Foodborne Diseases Active Surveillance Network) division of the CDC affirms in its 2017 report that the incidence of infections per 100,000 people was highest for *Campylobacter spp* and *Salmonella spp* from 2016 to 2017, which is similar to previous years. The situation in Europe for *Campylobacter spp* infections is illustrated by the Campylobacteriosis-Annual epidemiological report published the 30 January of 2017 by the ECDC (European Center of Disease Control) from data collected in 2014 [4]. The report affirms that 240,379 confirmed cases were reported in 2014 with a rate of campylobacteriosis of 59.8 cases per 100,000 population in the EU/EEA, representing a 13% increase compared with the previous year. Human campylobacteriosis was more common in children below five years of age and in general was slightly higher for males than females across all age groups. Campylobacteriosis shows a clear seasonality, with a sharp peak of cases in July, trend that it is confirmed by the CDC as well. At the beginning of the summer of 2018 in Pescara (Italy), 180 cases of intoxicated children were reported and identified as campylobacteriosis infection.

The most representative etiologic agent for campylobacteriosis is *Campylobacter jejuni*. It is a slender, spirally curved rod that possesses a single polar flagellum at one or both ends of the cell. It is oxidase and catalase positive, is microaerophilic, requiring small amounts of oxygen (3–6%) for growth; its optimum growth temperatures on solid media are 37 °C, and it grows well at pH 5.5–8.0. *C. jejuni* is associated with warm-blooded animals, in fact a large percentage of all major livestock animals have been shown to contain these organisms in their feces [5,6]. Some strains of *C. jejuni* produce a thermolabile enterotoxin (CJT), that has been reported to have some similar properties with the enterotoxins of *Vibrio cholerae* (CT) and *Escherichia coli* (LT) [7,8].

Many livestock warm-blooded animals can carry in their intestines, liver and giblets cells of *C. jejuni* that can be transferred to other edible parts of an animal when it's slaughtered. In the USA in 2014, National Antimicrobial Resistance Monitoring System (NARMS) testing found *C. jejuni* on 33% of raw chicken bought from retailers [9]. *Campylobacter spp* infection can also be transmitted through unpasteurized milk ingestion when a cow has *C. jejuni* cells in its udder or when milk is contaminated with manure [10]. Moreover, most problematic foods regarding *Campylobacter spp* infection are those consumed raw, such as fruits and vegetables that can be contaminated through contact with soil containing feces from cows, birds or other animals [11]. Animal feces can also contaminate water sources, such as lakes and streams.

Today there are many techniques that can be used in the identification of this type of contamination, many of which have limits related to the collection time of responses, high complexity or the possibility of being reused several times.

In the last few years, different kinds of sensors have been developed for *C. jejuni* detection. They are essentially DNA-biosensors, of which specificity is due to oligonucleotide probes covalently immobilized on the sensing surface. Several optical [12], acoustic [13] and electrochemical [14] techniques have been proposed for traducing the hybridization with the specific target nucleic acid to the pathogen detection [15]. As an example, quartz crystal microbalance (QCM) immunosensors, using monoclonal and polyclonal antibody systems coupled with the use of gold nanoparticles (AuNPs) to increase the sensitivity, were used [16]. In this way, a limit of detection (LOD) of 10^4 CFU/mL was reached; however, this kind of sensor is limited to a single use and consequently does not have a low usage cost. The same LOD was reached using a colorimetric aptasensor, that can be used for on-line applications and gives its response in 30 min [17].

As an example, to overcome time consumption and single-use limitations, approaches based on nanowire gas sensor technology could be employed. Nanowire gas sensors base their action principle on the analysis and recognition of the volatile fingerprint of a determinate sample. This kind of approach has already been applied successfully in many different fields as human microbiota monitoring [18], and environmental monitoring [19,20]. In particular, regarding food microbial contamination, nanowire tech was able to recognize the presence of a determinate microorganism throughout the set of volatile organic compounds (VOCs) emitted when growing in a determinate matrix [21–23]. In comparison with the aforementioned sensor technologies, nanowire gas sensors exhibit advantages of nanostructured materials such as long-term stability for sustained operations, high rate surface/area, drastically reduced time of response and the possibility to be reused as well. In this study, an array of these sensors has been used inside a portable device called Small Sensor System (S3), described in detail in Section 2.3.

The aim of this work was to find and identify the VOCs set that characterizes *C. jejuni* through Gas-Chromatography Mass-Spectrometry (GC-MS) and to assess the capability of S3 to distinguish between samples inoculated with this microorganism and control specimens to follow *C. jejuni* temporal development. The success of this study can lay the foundation to deepen the research in this field in order to use S3 as a tool for prevention of illness and food quality control in the future.

2. Materials and Methods

2.1. Samples Preparation

Samples were prepared using *C. jejuni* subsp. *jejuni* type strain purchased from DMSZ, DSM number 4688, (ATCC 33560, CCUG 11284, CIP 702, NCTC 11351) and Brain Heart Infusion Broth (BHI) media purchased from Sigma Aldrich (Merck). Tubes containing 9 mL of sterile BHI were inoculated with *C. jejuni* cells and were aerobically incubated for 24 h at 35 °C in order to produce enough biomass to proceed with the next step of analysis. After the incubation, the culture was used to inoculate tubes of sterile BHI media in order to realize the same optical density (OD) of the number 3 standard of McFarland that corresponds to a concentration of 9×10^8 CFU (Colony Forming Unit) by mL [24]. Subsequently, serial dilutions using sterile BHI media were performed until the concentration was decreased by 4 orders of magnitude to 9×10^4 CFU/mL. This concentration was used for the inoculation of the analyzed vials.

Sterile chromatography 20 mL vials containing 4 mL of BHI were inoculated with 100 μL of the 9×10^4 CFU/mL solution reaching a final concentration of 2.20×10^3 CFU/mL. Control vials were performed as well keeping the vials containing 4 mL of BHI with no inoculum. Furthermore, in order to control the effective number of cells at the beginning and the end of the analysis, a plate count technique was applied using four plates for each time (T0 and T20).

2.2. GC-MS Analysis

The Gas Chromatograph (GC) used during the analyses was a Shimadzu GC2010 PLUS (Kyoto, KYT, Japan), equipped with a Shimadzu single quadrupole Mass Spectrometer (MS) MS-QP2010 Ultra (Kyoto, KYT, Japan) and an autosampler HT280T (HTA s.r.l., Brescia, Italy). The GC-MS analysis was coupled with the Solid-Phase Micro Extraction (SPME) method in order to find the most characteristic VOCs for each sample.

The fiber used for the adsorption of volatiles was a DVB/CAR/PDMS-50/30 μm (Supelco Co., Bellefonte, PA, USA). The fiber was exposed to the headspace of the vials after heating and shaking the samples in the HT280T oven, thermostatically regulated at 40 °C for 15 min, with the aim of creating the headspace equilibrium. The length of the fiber in the headspace was kept constant. Desorption of volatiles took place in the injector of the GC-MS for 6 min at 240 °C.

The gas chromatograph was operated in the direct mode throughout the run, with the mass spectrometer in electron ionization (EI) mode (70 eV). GC separation was performed on a MEGA-WAX capillary column (30 m × 0.25 mm × 0.25 μm, Agilent Technologies, Santa Clara, CA, USA). Ultrapure helium (99.99%) was used as the carrier gas, at the constant flow rate of 1.5 mL/min. The following GC oven temperature programming was applied. At the beginning, the column was held at 40 °C for 3.5 min, and then raised from 40 to 90 °C at 5 °C/min. Next, the temperature was raised from 90 °C to 220 °C, with a rate of 12 °C/min; finally, 220 °C was maintained for 7 min.

The GC-MS interface was kept at 200 °C. The mass spectra were collected over the range of 40 to 500 m/z in the Total Ion Current (TIC) mode, with scan intervals at 0.3 s. The identification of the volatile compounds was carried out using the NIST11 and the FFNSC2 libraries of mass spectra.

Four samples were analyzed: control and *C. jejuni* at times T0 and T20.

2.3. S3 Analysis

The S3 device used in the present work has been completely designed and constructed at SENSOR Laboratory (University of Brescia, Brescia, Italy) in collaboration with NASYS S.r.l., a spin-off of the University of Brescia, Brescia, Italy. It has been described in other works [25–28]. Briefly, the tool comprises three parts: Pneumatic components that transfer VOCs from the headspace of samples to the sensing chamber; electronic boards that manage the acquisition and transmission of the data from the device to the dedicated Web-App and allow the synchronization between S3 and the auto-sampler; and a sensing chamber, that can host from five to ten different metal oxide (MOX) gas sensors and is thermostatically isolated in order to avoid any influence of the surrounding environment. To function properly, sensors need a reference value that has been obtained by filtering the ambient air with a small metal cylinder (21.5 cm in length, 5 cm in diameter) filled with activated carbons.

Eight MOX gas sensors were used. Three of them are MOX nanowire [29,30]. Two of them are tin oxide nanowires sensors, both grown by means of the Vapor Liquid Solid technique [31], using a gold catalyst on the alumina substrate and functionalizing one of them with gold clusters; the third sensor has an active layer of copper oxide nanowires. The working temperature is 350 °C, 350 °C and 400 °C, respectively. The other three sensors are prepared with Rheotaxial Growth and Thermal Oxidation (RGTO) thin film technology [32]; one is tin oxide functionalized with gold clusters (working at 400 °C), while the other two are pure tin oxide (working at 300 °C and at 400 °C, respectively). The last two are commercial MOX sensors produced by Figaro Engineering Inc. (Osaka, Japan). In particular, they are the TGS2611 and TGS2602, which are sensitive to natural gases and odorous gases like ammonia, respectively, according to the datasheet of the company. Commercial sensors have been mounted on our e-nose in order to evaluate the performances of nanowire sensors. Details of S3 sensors made at SENSOR Laboratory are summarized in Table 1. Response to 5 ppm of ethanol, selectivity (response ethanol/response carbon monoxide) and limit of detection (LOD) of ethanol are also included.

Table 1. Type, composition, morphology, operating temperature, response ($\Delta R/R$), selectivity (response ethanol/response carbon monoxide) and limit of detection (LOD) of ethanol for S3 sensors made at the SENSOR Laboratory.

Materials (Type)	Composition	Morphology	Operating Temperature (°C)	Response to 5 ppm of Ethanol	Selectivity	Limit of Detection (LOD) of Ethanol (ppm)
SnO_2Au (n)	SnO_2 functionalized with Au clusters	RGTO	400 °C	6.5	3	0.5
SnO_2 (n)	SnO_2	RGTO	300 °C	3.5	2.5	1
SnO_2 (n)	SnO_2	RGTO	400 °C	4	2	0.8
SnO_2Au + Au (n)	SnO_2 grown with Au and functionalized with gold clusters	Nanowire	350 °C	7	2.5	0.5
SnO_2Au (n)	SnO_2 grown with Au	Nanowire	350 °C	5	2.1	1
CuO (p)	CuO	Nanowire	400 °C	1.5	1.5	1

The autosampler headspace system HT2010H (HTA s.r.l., Brescia, Italy) was coupled with S3. It supports a 42-loading-sites carousel and a shaking oven to equilibrate the sample headspace. 40 vials were placed in a randomized mode into the carousel. Among these vials, 5 were control samples analyzed at times 2.5 h, 5.5 h, 8.5 h, 11.5 h and 14.5 h. Each vial was incubated at 40 °C for 10 min in the autosampler oven. The sample headspace was then extracted from the vial in the dynamic headspace path and released into the carried flow (100 sccm). The analysis timeline can be divided into three different steps for a duration of 30 min per sample divided as follows: 5 min to analyze samples, 5 min to clean the tube that carries VOCs from sample headspace to sensing chamber and 20 min to restore the baseline. Thanks to the processor integrated in the S3 instrument, the frequency at which the equipment works is equal to 1 Hz.

2.4. S3 Data Analysis

Data analysis was performed using MATLAB® R2015a software (MathWorks, Natick, MA, USA). First of all, sensors responses in terms of resistance (Ω) were normalized when compared to the first value of the acquisition (R_0). For all the sensors, the difference between the first value and the minimum value during the analysis time was calculated; hence, $\Delta R/R_0$ has been extracted as featured.

Principal Component Analysis (PCA) was applied to this data matrix in order to evaluate the ability of the system to follow variation over time of the number of bacteria and therefore also the quantity of VOCs emitted.

3. Results and Discussion

3.1. GC-MS Results

Chromatograms of analyzed samples were compared to highlight differences between control samples and *C. jejuni* ones to see the variation of emitted VOCs in the headspace between T0 and T20.

The comparison between control and *C. jejuni* specimens at time T0 underlines no significant differences in terms of number and amount of VOCs. In Table 2, the list of compounds with their retention time (RT) and abundance in arbitrary unit is reported. Correlation coefficient has been calculated to get how similar the two samples were; a value equal to 0.9965 has been obtained. This proves that during the conditioning period of 15 min before fiber exposure in the GC injector, *C. jejuni* VOCs production was too small to change headspace composition.

Table 2. List of volatile organic compounds (VOCs) for *C. jejuni* and control samples with their retention times (RT) and abundance in arbitrary units at time T0.

RT	VOC	Abundance	
		Campylobacter	Control
1.552	3-Butynol	5,488,041	5,289,186
2.674	Isovaleraldehyde	28,336,125	25,535,856
5.386	Dimethyl Disulfide	5,401,037	6,048,406
8.666	3-*O*-Methyl-D-fructose	912,105	904,062
9.281	1-Hydroperoxyheptane	533,178	418,838
12.332	2,5-Dimethylpyrazine	623,139	560,597
14.432	Nonanal	512,867	227,918
14.624	6-Methyloctadecane	549,617	794,739
15.419	4-Methyl-2-oxovaleric acid	457,664	412,957
15.805	2-Acetylamino-3-hydroxy-propionic acid	27,861	51,993
16.017	1-(2-Methoxy-1-methylethoxy)-2-propanol	519,216	240,149
16.372	Ethylhexanol	404,819	389,367
16.866	Benzaldehyde	12,509,648	13,515,034
17.430	3-Trifluoroacetoxydodecane	92,701	168,299
18.455	3-Hydroxycyclohexanone	145,875	206,891
18.695	Acetophenone	2,201,980	1,987,623
19.545	[(2-Ethylhexyl)methyl]oxirane	93,105	281,168
19.950	Methoxy-phenyl-oxime	1,545,714	1,301,333
20.871	Heptanoic acid	268,343	304,130
21.205	Benzyl alcohol	280,516	246,860
21.555	2-[2-(Benzyloxy)-1-(1-methoxy-1-methylethoxy)ethyl]oxirane	226,967	202,114
22.021	1-Dodecanol	222,887	482,480
22.465	Phenyl carbamate	51,364	93,104
23.580	Octanal	121,544	130,790
24.480	Octadecanal	131,841	76,150
25.078	2,6-Bis(tert-butyl)phenol	691,529	597,086
27.565	N,N-Dimethylformamide ethylene acetal	53,073	24,830

On the contrary, sample analysis after 20 h has indicated changes in vial headspace due to microorganisms metabolic activity and to slow release of VOCs contained in BHI broth. In Table 3, the list of VOCs is shown. Main differences between the specimens reside in the presence of alcohol compounds, such as 1-pentanol, acetoin, 2,7-dimethyl-4,5-octanediol, 2-propyl-1-pentanol, bicyclo[3.2.1]octan-6-ol, 1-nonanol, γ-methylmercaptopropyl alcohol and (9*E*)-9-hexadecen-1-ol greater in *C. jejuni* samples than control one. This result points out how sugar fermentation went on during a 20 h incubation period at 37 °C. Furthermore, this heating phase could also be responsible for the formation of new compounds derived from pyrazine, like trimethylpyrazine and

2-ethyl-3,6-dimethylpyrazine. In this case, the correlation coefficient is equal to 0.2666, indicating the samples were strongly diverse.

Table 3. List of VOCs for *C. jejuni* and control samples with their retention times (RT) and abundance in arbitrary units at time T20.

RT	VOC	Abundance	
		Campylobacter	Control
1.540	3-Butynol	8,834,835	2,135,579
2.266	Isovaleraldehyde	221,889,809	3,852,176
4.142	Dimethyl Disulfide	42,167,678	0
8.377	1-Pentanol	236,112,172	0
9.229	Isoamyl Alcohol	0	737,186
10.673	Acetoin	1,974,910	0
11.125	2-Methylbutyl isovalerate	1,684,776	0
12.299	2,5-Dimethylpyrazine	0	693,770
14.400	Nonanal	0	123,487
14.483	Trimethylpyrazine	0	258,896
14.600	6-Methyloctadecane	0	60,501
15.275	2-Ethyl-3,6-dimethylpyrazine	243,733	0
15.392	4-Methyl-2-oxovaleric acid	454,870	273,645
15.702	Ammonium acetate	812,172	572,057
15.903	2,7-Dimethyl-4,5-octanediol	945,696	0
16.186	1-(2-Methoxy-1-methylethoxy)-2-propanol	0	79,930
16.295	2-Propyl-1-pentanol	308,880	0
16.340	Ethylhexanol	267,747	301,354
16.845	Benzaldehyde	0	14,760,479
17.366	1-Octanol	972,272	256,675
17.550	Bicyclo[3.2.1]octan-6-ol	164,638	0
17.946	2-Undecanone	103,647	0
18.419	3-Hydroxycyclohexanone	0	79,109
18.581	Benzeneacetaldehyde	5,087,169	0
18.668	Acetophenone	0	669,190
18.740	1-Nonanol	1,331,499	0
18.914	Methyl 4-hydroxybutanoate	0	204,876
19.446	γ-Methylmercaptopropyl alcohol	690,884	0
19.923	E-11,13-Tetradecadien-1-ol	2,393,731	532,815
20.575	β-Phenethyl acetate	131,503	0
20.844	Heptanoic acid	488,771	215,935
21.187	Benzyl alcohol	342,619	146,035
21.540	Phenylethyl Alcohol	26,577,900	1,429,647
21.750	m-Tolunitrile	0	60,783
21.996	1-Dodecanol	492,208	270,053
22.316	Tropone	165,612	56,068
22.443	4-Hydroxybenzenephosphonic acid	0	77,574
22.684	Nerolidyl acetate	0	116,357
22.876	Octanoic acid	178,530	73,913
23.555	1,3,2-Dioxaborolane, 2-ethyl-4-(3-oxiranylpropyl)-	0	48,973
23.822	(9*E*)-9-Hexadecen-1-ol	188,001	0
25.047	2,4-Di-tert-butylphenol	204,650	200,860
26.632	Pyrindan	103,007,890	14,885,449
27.555	N,N-Dimethylformamide ethylene acetal	40,116	40,546

Growth of *C. jejuni* is also confirmed by microbiological analysis. There were $(8.57 \pm 1.18) \times 10^4$ CFU/mL at time T0 in terms of mean $\pm$ standard deviation calculated on four plates and $(1.38 \pm 0.40) \times 10^7$ CFU/mL at time T20.

3.2. S3 Results

First step of S3 data analysis consisted of checking which of the eight sensors were performing more. Sensor responses were normalized in order to highlight the variation of the resistance once sensing materials were exposed to VOCs. Five sensors showed the best performances: two RGTO (SnO_2Au and SnO_2 heated at 300 °C), SnO_2Au nanowire, copper oxide and TGS2611. Resistance variation over time for all measures is shown in Figure 1 for three kinds of sensors, i.e., RGTO, nanowire and commercial MOX.

CuO sensing material exhibits an increase in resistance with respect to the R_0 value, while TGS2611 and SnO_2 RGTO have an opposite behavior due to their n-type semiconductor characteristic. However, all of them are characterized by the decreasing of ΔR with the growth of time. This trend is more evident for CuO and TGS2611 sensors. Conversely, RGTO has an increase of ΔR for the first five samples, while from the seventh specimen it has the same kind of response as the other two, even if it is less accentuated. This tendency could be explained considering that the number of microorganisms grows over time very quickly; it has been shown that they double their number in BHI broth in 75 min (average value) [33]. Hence, many VOCs could be used by *C. jejuni* to feed, thus subtracting them from the headspace. At the same time, new compounds are emitted from bacteria and pass in the gaseous phase, as shown in Tables 2 and 3. It is important to underline that for 20 h not only the alcohols have increased in number and amount, but also other compounds. Among them, the one present in greater quantity is pyrindan, a bicyclic compound containing a pyridine ring. The reduction of $\Delta R/R_0$ value could be due to action of this compound. Furthermore, it can be noticed that the first sample of the analysis produced a ΔR significantly different in respect to the others; it is higher for all sensors, especially for RGTO and nanowire sensors. Since this could be the result of a different conditioning, it has been discarded for the following analysis.

Figure 2 refers to PCA that has been performed using the five aforementioned sensors. The first two Principal Components (PC) were used for a total explained variance equal to 99.08% (91.77% in PC1 and 7.31 in PC2). It is possible to identify two different trends. For *C. jejuni* samples, first four samples (0.5–2.5 h) are characterized by descending scores along PC2 axis, while the other specimens (3–20 h) are distributed essentially along PC1 ascending scores. Furthermore, the distance between points decreases as time goes on and it can be explained by the typical growth curve of microorganism. Indeed, it is characterized by four phases: (A) lag phase (bacteria adapt themselves to growth conditions and are yet not able to divide), (B) log phase (cell doubling), (C) stationary phase (growth rate and stationary rate are equal due to a growth-limiting factor such as the depletion of an essential nutrient) and (D) death phase (bacteria die). In PCA, lag phase corresponds to the first four points that move along PC2 axis, log phase to the following eleven samples (3–8 h) and stationary phase to the remaining ones (8.5–20 h). Instead, control samples assume higher scores along both PCs axes with increasing time. It is due to the slow release in the headspace of some VOCs contained in the BHI broth. The only exception is the fourth control sample that does not follow the parabolic trend of the others, but it is closer to *C. jejuni* points.

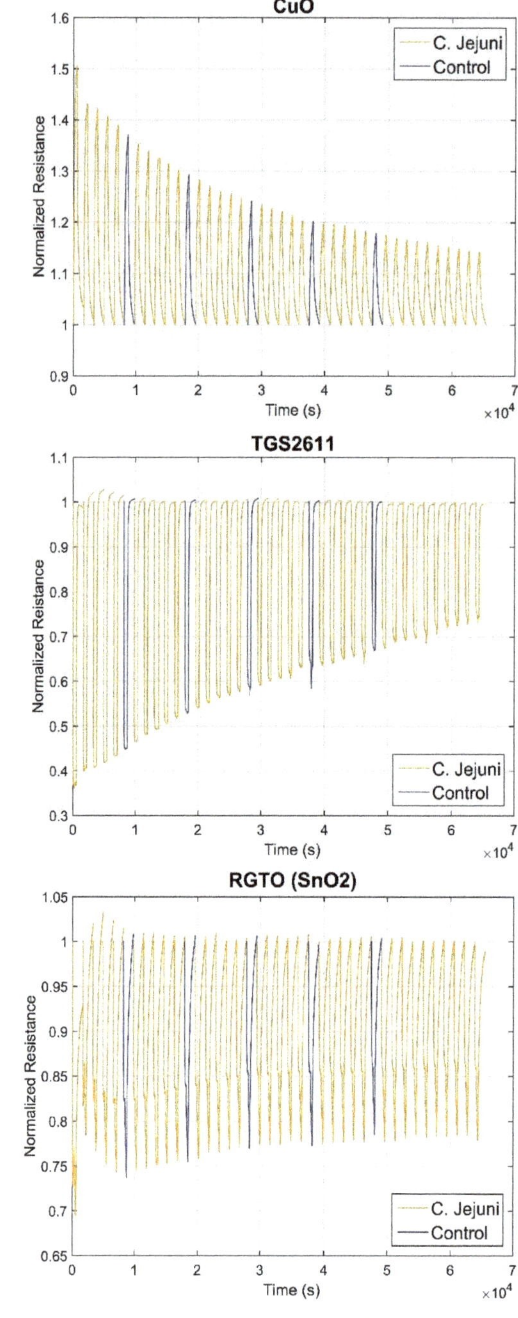

Figure 1. Resistance variations of three sensors once exposed to VOCs. From top to bottom: copper oxide nanowire, TGS2611 and tin oxide Rheotaxial Growth and Thermal Oxidation (RGTO). On the *x*-axis there is time in seconds, on the *y*-axis normalized resistance.

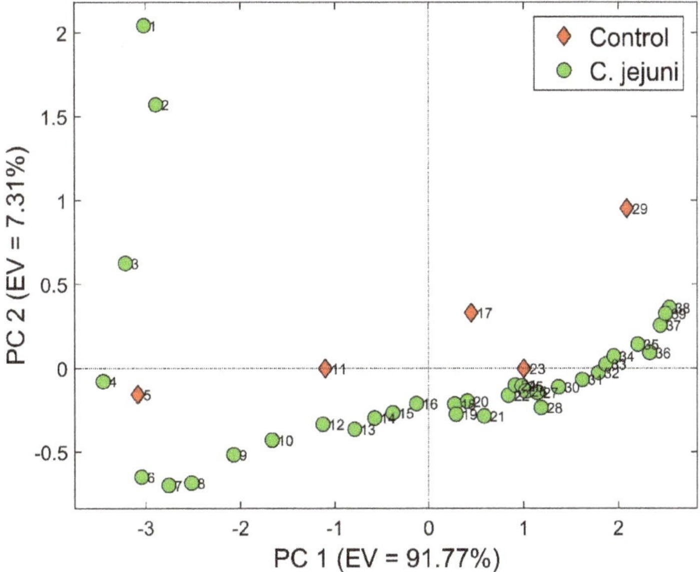

Figure 2. Principal Component Analysis (PCA) done with first two components (total variance equal to 99.08%). Green circles are *C. jejuni* samples, red diamond control ones.

4. Conclusions

This work demonstrates the potential of this technology for the development of a device able to give a response very quickly when compared to classical microbiological methods, 5 min for the former and several hours for the latter. This lays the foundation to develop a rapid and sensitive detection method for *C. jejuni* in the future. Innovative gas sensors with nanowire and RGTO morphologies used in this study have proven to be useful tools for the identification and characterization of microbiological contamination. In particular, PCA done with five sensors shows the capability of the system to follow bacteria growth along a period of 20 h. During which the sensors used were able to associate the response faithfully following the growth curve of the contaminated microorganisms. We will plan to continue this study by focusing on reducing the detection threshold in order to use this tool to individuate the presence of *C. jejuni* at low concentrations and to avoid human infections. We will evaluate in future works how the food matrix where *C. jejuni* develops and grows will influence sensor responses and their LODs.

Author Contributions: Conceptualization, E.N.-C; Methodology, E.N.-C.; software, M.A.; Validation, E.N.-C., M.A. and V.S.; Formal analysis, M.A.; Data curation, M.A.; Writing—original draft preparation, E.N.-C., M.A. and V.S.; Writing—review and editing, E.N.-C., M.A. and V.S.; Supervision, V.S.

Funding: This research received no external funding.

References

1. Yeung, R.M.W.; Morris, J. Food safety risk—Consumer perception and purchase behavior. *Br. Food J.* **2001**, *103*, 170–187. [CrossRef]
2. Yu, H.; Neal, J.A.; Sirsat, S.A. Consumers' food safety risk perceptions and willingness to pay for fresh-cut produce with lower risk of foodborne illness. *Food Control* **2018**, *86*, 83–89. [CrossRef]
3. Centers for Disease Control and Prevention, Foodborne Illnesses and Germs. Available online: https://www.cdc.gov/foodsafety/foodborne-germs.html (accessed on 31 October 2018).

4. European Centers for Disease Prevention and Control, Campylobacteriosis—Annual Epidemiological Report 2016. Available online: https://ecdc.europa.eu/en/publications-data/campylobacteriosis-annual-epidemiological-report-2016-2014-data (accessed on 31 October 2018).

5. Açik, M.N.; Çetinkaya, B. Heterogeneity of Campylobacter jejuni and Campylobacter coli strains from healthy sheep. *Vet. Microbiol.* **2006**, *115*, 370–375. [CrossRef] [PubMed]

6. Brown, P.E.; Christensen, O.F.; Clough, H.E.; Diggle, P.J.; Hart, C.A.; Hazel, S.; Kemp, R.; Leatherbarrow, A.J.H.; Moore, A.; Sutherst, J.; et al. Frequency and spatial distribution of environmental Campylobacter spp. *Appl. Environ. Microbiol.* **2004**, *70*, 6501–6511. [CrossRef]

7. Klipstein, F.A.; Engert, R.F. Properties of Crude Campylobacter jejuni Heat-Labile Enterotoxin. *Infect. Immun.* **1984**, *45*, 314–319. [PubMed]

8. Jay, J.M.; Loessner, M.J.; Golden, D.A. Foodborne Gastroenteritis Caused by *Vibrio*, *Yersinia* and *Campylobacter*. In *ModernFood Microbiology*, 7th ed.; Heldman, D.R., Ed.; Heldman Associates: San Marcos, CA, USA, 2005; Volume 1, pp. 657–678, ISBN 0-387-23180-3.

9. The National Antimicrobial Resistance Monitoring System: Enteric Bacteria. Available online: https://www.google.com/url?q=https://www.fda.gov/downloads/AnimalVeterinary/SafetyHealth/AntimicrobialResistance/NationalAntimicrobialResistanceMonitoringSystem/UCM528861.pdf&sa=D&ust=1543574350805000&usg=AFQjCNEMaE6VwxnVNmUYIOqrV7HnmBpcNQ (accessed on 30 November 2018).

10. Artursson, K.; Schelin, J.; Lambertz, S.T.; Hansson, I.; Engvall, E.O. Foodborne pathogens in unpasteurized milk in Sweden. *Int. J. Food Microbiol.* **2018**, *284*, 120–127. [CrossRef] [PubMed]

11. Verhoeff-Bakkenesa, L.; Jansenb, H.A.P.M.; In't Veldb, P.H.; Beumerc, R.R.; Zwieteringc, M.H.; van Leusden, F.M. Consumption of raw vegetables and fruits: A risk factor for Campylobacter infections. *Int. J. Food Microbiol.* **2011**, *144*, 406–412. [CrossRef]

12. Cecchini, F.; Manzano, M.; Mandabi, Y.; Perelman, E.; Marks, R.S. Chemiluminescent DNA optical fibre sensor for Brettanomyces bruxellensis detection. *J. Biotechnol.* **2012**, *157*, 25–30. [CrossRef] [PubMed]

13. Jia, K.; Toury, T.; Ionescu, R.E. Fabrication of an atrazine acoustic immunosensor based on a drop-deposition procedure. *IEEE Trans. Ultrason. Ferroelectr. Freq. Control* **2012**, *59*, 2015–2021.

14. Kubičárová, T.; Fojta, M.; Vidic, J.; Tomschik, M.; Suznjevic, D.; Paleček, E. Voltammetric and chronopotentiometric measurements with nucleic acid-modified mercury film on a glassy carbon electrode. *Electroanalysis* **2000**, *12*, 1390–1396. [CrossRef]

15. Vidic, J.; Manzano, M.; Chang, C.M.; Jaffrezic-Renault, N. Advanced biosensors for detection of pathogens related to livestock and poultry. *Vet. Res.* **2017**, *48*, 1–22. [CrossRef] [PubMed]

16. Masdor, N.A.; Altintas, Z.; Tothilla, I.E. Sensitive detection of Campylobacter jejuni using nanoparticles enhanced QCM sensor. *Biosens. Bioelectron.* **2016**, *78*, 328–336. [CrossRef] [PubMed]

17. Kim, Y.J.; Kim, H.S.; Chon, J.W.; Kim, D.H.; Hyeon, J.Y.; Seo, K.H. New colorimetric aptasensor for rapid on-site detection of Campylobacter jejuni and Campylobacter coli in chicken carcass samples. *Anal. Chim. Acta* **2018**, *1029*, 78–85. [CrossRef] [PubMed]

18. Sberveglieri, V.; Núñez-Carmona, E.; Ponzoni, A.; Comini, E.; Galstyan, V.; Zappa, D.; Pulvirenti, A. Skin microbiota monitoring by nanowire MOS sensors. In Proceedings of the 29th European Conference on Solid-State Transducers, EUROSENSORS 2015, Freiburg, Germany, 6–9 September 2015; Volume 120, pp. 756–759.

19. Núñez-Carmona, E.; Soprani, M.; Sberveglieri, V. Nanowire (S3) Device for the quality control of drinking water. In *Smart Sensors, Measurement and Instrumentation*; Mukhopadhyay, S.C., Ed.; Springer: Berlin/Heidelberg, Germany, 2007; Volume 23, pp. 179–203.

20. Núñez-Carmona, E.; Sberveglieri, V.; Ponzoni, A.; Zappa, D.; Pulvirenti, A. Small Sensor Sistem S3 device to control the microbial contamination in water. In Proceedings of the 9th International Conference on Sensing Technology (ICST), Auckland, New Zealand, 8–11 December 2015; pp. 246–250.

21. Núñez-Carmona, E.; Sberveglieri, V.; Ponzoni, A.; Galstyan, V.; Zappa, D.; Pulvirenti, A.; Comini, E. Detection of food and skin pathogen microbiota by means of an electronic nose based on metal oxide chemiresistors. *Sens. Actuators B-Chem.* **2017**, *238*, 1224–1230. [CrossRef]

22. Zambotti, G.; Sberveglieri, V.; Gobbi, E.; Falasconi, M.; Núñez-Carmona, E.; Pulvirenti, A. Fast identification of microbiological contamination in vegetable soup by electronic nose. In Proceedings of the 28th European Conference on Solid-State Transducers, EUROSENSORS 2014, Brescia, Italy, 7–10 September 2014; pp. 246–250.
23. Ponzoni, A.; Baratto, C.; Cattabiani, N.; Falasconi, M.; Galstyan, V.; Núñez-Carmona, E.; Rigoni, F.; Sberveglieri, V.; Zambotti, G.; Zappa, D. Smetal oxide gas sensors, a survey of selectivity issues addressed at the SENSOR lab, Brescia (Italy). *Sensors* **2017**, *17*, 714. [CrossRef]
24. McFarland Standards. Available online: http://www.dalynn.com/dyn/ck_assets/files/tech/TM53.pdf (accessed on 30 November 2018).
25. Abbatangelo, M.; Núñez-Carmona, E.; Sberveglieri, V.; Zappa, D.; Comini, E.; Sberveglieri, G. Application of a Novel S3 Nanowire Gas Sensor Device in Parallel with GC-MS for the Identification of Rind Percentage of Grated Parmigiano Reggiano. *Sensors* **2018**, *18*, 617. [CrossRef] [PubMed]
26. Abbatangelo, M.; Núñez-Carmona, E.; Sberveglieri, V. Application of a novel S3 nanowire gas sensor device in parallel with GC-MS for the identification of Parmigiano Reggiano from US and European competitors. *J. Food Eng.* **2018**, *236*, 36–43. [CrossRef]
27. Sberveglieri, V. Validation of Parmigiano Reggiano Cheese Aroma Authenticity, Categorized through the Use of an Array of Semiconductors Nanowire Device (S3). *Materials* **2016**, *9*, 81. [CrossRef]
28. Sberveglieri, V.; Bhandari, M.P.; Núñez-Carmona, E.; Betto, G.; Sberveglieri, G. A novel MOS nanowire gas sensor device (S3) and GC-MS-based approach for the characterization of grated Parmigiano Reggiano cheese. *Biosensors* **2016**, *6*, 60. [CrossRef]
29. Ponzoni, A.; Zappa, D.; Comini, E.; Sberveglieri, V.; Faglia, G.; Sberveglieri, G. Metal oxide nanowire gas sensors: Application of conductometric and surface ionization architectures. *Chem. Eng. Trans.* **2012**, *30*, 31–36. [CrossRef]
30. Sberveglieri, G.; Concina, I.; Comini, E.; Falasconi, M.; Ferroni, M.; Sberveglieri, V. Synthesis and integration of tin oxide nanowires into an electronic nose. *Vacuum* **2012**, *86*, 532–535. [CrossRef]
31. Wagner, R.S.; Ellis, W.C. Vapor-Liquid-Solid Mechanism of Single Crystal Growth. *Appl. Phys. Lett.* **1964**, *4*, 89–90. [CrossRef]
32. Sberveglieri, G. Recent developments in semiconducting thin-film gas sensors. *Sens. Actuators B-Chem.* **1995**, *23*, 103–109. [CrossRef]
33. Battersby, T.; Walsh, D.; Whyte, P.; Bolton, D.J. Campylobacter growth rates in four different matrices: Broiler caecal material, live birds, Bolton broth, and brain heart infusion broth. *Infect. Ecol. Epidemiol.* **2016**, *6*. [CrossRef] [PubMed]

 *biosensors*

Article

ZIF Nanocrystal-Based Surface Acoustic Wave (SAW) Electronic Nose to Detect Diabetes in Human Breath

Fabio A. Bahos [1], Arianee Sainz-Vidal [1], Celia Sánchez-Pérez [1], José M. Saniger [1], Isabel Gràcia [2], María M. Saniger-Alba [3] and Daniel Matatagui [1,4,*]

[1] Instituto de Ciencias Aplicadas y Tecnología (ICAT), Universidad Nacional Autónoma de México, Ciudad Universitaria, Ciudad de México 04510, Mexico; fbahos@gmail.com (F.A.B.); arianee.sainz@ccadet.unam.mx (A.S.-V.); celia.sanchez@ccadet.unam.mx (C.S.-P.); jose.saniger@ccadet.unam.mx (J.M.S.)

[2] Instituto de Microelectrónica de Barcelona (IMB), CSIC, Campus UAB, 08193 Bellaterra, Spain; isabel.gracia@imb-cnm.csic.es

[3] Instituto Nacional de la Nutrición Salvador Zubiran, Department of Neurophysiology, Tlalpan 14080, Mexico; mariadelmarsaniger@gmail.com

[4] SENSAVAN, Instituto de Tecnologías Físicas y de la Información (ITEFI), CSIC, Serrano 144, 28006 Madrid, Spain

* Correspondence: d.m@csic.es; Tel.: +34-91-561-88-06 (ext. 920422)

Received: 7 November 2018; Accepted: 21 December 2018; Published: 26 December 2018

Abstract: In the present work, a novel, portable and innovative eNose composed of a surface acoustic wave (SAW) sensor array based on zeolitic imidazolate frameworks, ZIF-8 and ZIF-67 nanocrystals (pure and combined with gold nanoparticles), as sensitive layers has been tested as a non-invasive system to detect different disease markers, such as acetone, ethanol and ammonia, related to the diagnosis and control of diabetes mellitus through exhaled breath. The sensors have been prepared by spin coating, achieving continuous sensitive layers at the surface of the SAW device. Low concentrations (5 ppm, 10 ppm and 25 ppm) of the marker analytes were measured, obtaining high sensitivities, good reproducibility, short time response and fast signal recovery.

Keywords: eNose; gas sensor; SAW; surface acoustic wave; Love wave; diabetes; breath; VOC; ZIF; Zeolite

1. Introduction

One of the great challenges of contemporary science is the efficient diagnosis of diseases using non-invasive techniques. This strategy aims to provide a higher quality of life for humans and reduce the mortality rate. Additionally, early treatment of diseases and its complications has an important economic impact by helping to avoid or reduce treatment costs.

The last portion of deeply exhaled breath, representing alveolar air, can be considered the headspace gas of blood. Exhaled breath, recognized mainly through the sense of smell, is a method that has long been used for disease diagnosis. This method was abandoned due to the emergence of new accurate and effective techniques, despite being highly invasive. Over the last few decades, an important advance in gas analysis technologies has re-launched the idea of diagnosing diseases by analyzing exhaled breath. Various studies using these techniques, such as gas chromatography-mass spectrometry (GC-MS), proton transfer reaction-mass spectrometry (PTR-MS), selected ion flow tube-mass spectrometry (SIFT-MS), ion mobility and optical absorption [1–5], have shown a link between the chemical composition of exhaled breath and certain diseases. Chemical compounds present in exhaled breath that change due to diseases are known as markers. The conventional systems mentioned above are accurate but are also bulky, expensive, and require highly-qualified operators. This has created a demand for low-cost systems with high sensitivity and small dimensions based on solid-state chemical sensors, with different detection principles such as impedance [6], resistive [7,8], optical [9] and piezoelectric [10,11], surface acoustic wave (SAW), the last of which is one of the most sensitive piezoelectric devices [12]. The design of materials with advanced features led to a new generation of chemical sensors with enhanced sensitivity and response time [13–15]. Due to their unique porous structure zeolites have been used to detect gases [16]. However, in the last few years, organic zeolites such as zeolitic imidazolate frameworks (ZIFs) have attracted major attention as gas detectors [17–21], because they offer two primary advantages over conventional zeolites. First, they have larger pore sizes (about 1.16 nm for ZIF-8 and ZIF-67) and usually a smaller crystal size, resulting in higher surface area. Second, hydrophobic behavior is more pronounced in many ZIFs [22–24].

The World Health Organization (WHO) recognizes Diabetes mellitus, known as diabetes, as a serious and chronic disease that in 2012 caused 1.5 million deaths. A recent study reported in 2015 from 111 countries, estimated that there were 415 million people with diabetes aged 20–79 years, 5 million deaths attributable to diabetes, and the total global health expenditure due to diabetes was estimated to be 673 billion US dollars representing a substantial clinical and public health burden [8]. Moreover, the number of cases of diabetes among youths [25] and infancy [26] has increased in recent years but information on recent incidence trends is lacking and only statistical data for some countries are available.

The presence of ketones in exhaled breath is a warning sign of ketosis that is related to fat catabolism either due to carbohydrate deprivation or its lack of utilization in persons with diabetes. This condition is known as diabetic ketoacidosis and requires immediate treatment. One type of ketone, known as acetone, provides a non-invasive measure of ketosis through breath. The basal level of acetone in a healthy people can be around 2 ppm [5,27]. Adults following low-carbohydrate diets can have elevated levels of ketones up to 40 ppm [28–30], and poorly controlled diabetes can cause ketoacidosis which can increase acetone concentration up to 1250 ppm [27,31]. However, human exhaled air is a complex mixture of chemical compounds, making the detection and stage classification of a determinate disease through a unique marker difficult, so that different disease markers need to be considered as indicators. Another marker related to blood glucose concentration and present in exhaled breath is ethanol, which is not directly produced by any known mammalian cellular biochemical pathway, and may increase in exhaled gas mixtures because of alcoholic fermentation of an excessive overload of carbohydrate-rich food in conjunction with overgrowth of intestinal bacteria. Ethanol used in combination with exhaled acetone allowed the prediction of fluctuating plasma glucose concentrations in a multi-linear regression model [32–35], demonstrating that it can be helpful in determining diabetes through exhaled breath. However, diabetes is the cause of half of the cases of renal failure. Kidney failure is related with ammonia levels higher than 3 ppm in exhaled breath [5,30,36,37]. Consequently, a finger print using breath levels of acetone, ethanol and ammonia could be a non-invasive predictor of diabetes, its control, and a proxy for damage caused by the disease.

In the present work, a SAW eNose, based on ZIF nanocrystals as sensitive layers, has been tested to detect acetone, ethanol, and ammonia as a potential non-invasive system to diabetes diagnosis and control.

2. Materials and Methods

2.1. Materials

All the reagents were purchased from a commercial provider (Sigma-Aldrich, St. Louis, MO, USA) and used without further purification, including $ZnCl_2$ (98%), $CoCl_2·6H_2O$ (98%), and 2methylimidazole (99%), hydrogen tetrachloroaurate (III) hydrate ($HAuCl_4$, 99.9%) and trisodium citrate dihydrate (HOC (COONa) $(CH_2COONa)_2 · 2H_2O$).

2.2. Synthesis of ZIF-8 and ZIF-67

ZIF-8 and ZIF-67 samples were synthesized using the aqueous method reported elsewhere [38]. For ZIF-8 synthesis, a solution of 1.17 g of zinc chloride dissolved in 8 mL deionized (DI) water was added into a solution of 2-methylimidazole (2MeIM) (22.70 g) dissolved in 80 mL DI water, to yield a molar ratio of 2-methylimidazole to zinc of 70:1. The mixture was stirred at room temperature for 5 min. The product was collected by centrifugation (24,000 rpm, 10 min), washed in DI water three times and dried at 65 °C for 24 h in an oven. ZIF-67 was synthesized identically to the ZIF-8 material as described above, replacing zinc chloride with the equivalent quantity of cobalt chloride hexahydrate. A scheme of the synthesis paths is described in Figure 1 [39,40].

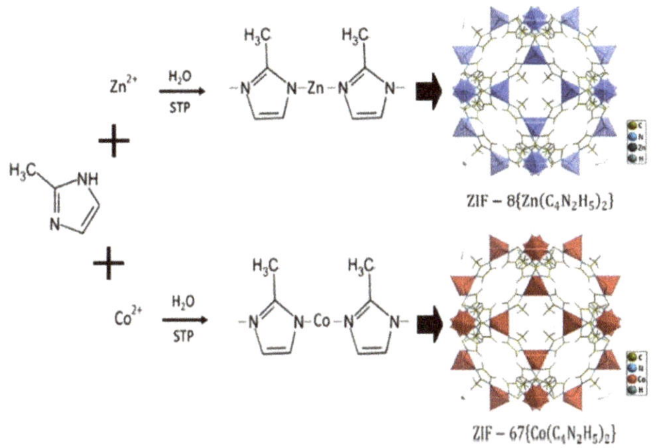

Figure 1. Representative synthesis and crystal structures of ZIF-8 and ZIF-67 used as the sensitive layers of gas sensors.

2.3. Synthesis of Gold Nanoparticles

Gold nanoparticles (AuNP) were prepared following procedure described in reference [41]. An aqueous solution of HAuCl$_4$ (0.001 M, 40 mL) was placed into a 250 mL round bottom flask. The solution was heated to 90 °C followed by the addition of sodium citrate aqueous solution (38.8 mM, 2 mL) into it while stirring for 15 min. After cooling down to room temperature, the solution was centrifuged three times with ethanol and three times with DI water, finally the precipitate was re-dispersed in DI water, resulting a water solution with AuNP of ~5 nm.

2.4. Zeolitic Imidazolate Framework Nanocrystal Characterization

The synthetized samples were kept at room conditions and characterized using: Fourier transform-infrared spectroscopy (FTIR), X-ray diffraction (XRD), scanning electron microscope (SEM) and energy-dispersive X-ray (EDS). FTIR spectra were recorded using a Thermo Nicolet NEXUS 670 FTIR spectrometer. The sample was diluted into KBr pellets in a 1:100 weight ratio (sample to KBr). The scanning range was 400–4000 cm^{-1} and the resolution was 4 cm^{-1}. XRD powder patterns were recorded with CuKα radiation in a D8 advance diffractometer from Bruker. The morphological features were examined by SEM. The SEM and EDS analysis were performed on a JEOL JMS-7600F.

2.5. Love-Wave Sensor

Love-waves (LW) are a specific type of SAW sensors based on shear horizontal (SH) waves guided by a layer with a lower propagation velocity than that of the piezoelectric substrate. The energy of the wave is confined in the guiding layer and any perturbation in it affects the acoustic wave velocity. The LW sensors used in the present work were designed with a delay line (DL) configuration. This device is based on a micro-electromechanical system composed of a piezoelectric material (ST-Quartz) with facing input/output aluminum interdigital transducers (IDT) on its surface, working at a 28 µm wavelength (λ), with a separation between IDTs of 2100 µm (Figure 2a). The SH waves were guided by a 3.1 µm thick layer of SiO_2, with the sensitive layer at its surface. An oscillator circuit consisting of a DL, with an amplification stage and a coupler were used for measuring the changes in the velocity of the waves by means of the resonant frequency (Figure 2b).

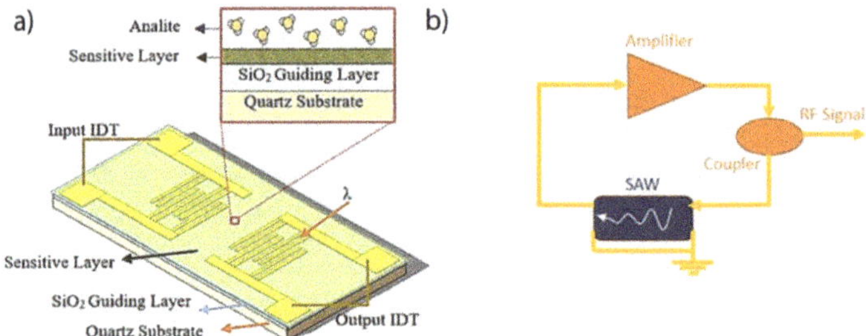

Figure 2. (**a**) Scheme representation of LW sensor and layer composition. (**b**) Oscillator circuit used to read the resonant frequency.

2.6. ZIF as Sensitive Layers

The eNose was based on a SAW sensor array with different sensitive layers to achieve a specific fingerprint for analytes of interest. The sensitive material samples were obtained by mixing each main solution with a volumetric proportion of 75% (solution-1) and 25% (solution-2) (Table 1). Spin coating was used to deposit a thin layer of the sensitive material. The process consisted of putting 25 µL of sample directly on the SiO_2 guiding layer, completely covering the IDTs surface and the area between them, and then spinning the assemblage at a speed of 3000 rpm for one minute, ensuring that the sensing area remained constant from sensor to sensor (Figure 3). A suitable thickness for each sensor is obtained after depositing four times in a multilayer configuration achieving an optimal sensitivity for sensors. After ZIFs deposition process a thermal treatment at 180 °C with 50 mL/min nitrogen flow in a tubular oven was applied during four hours for extracting the excess of the 2MeIM organic ligand in the sensitive layer.

Table 1. Different composition of the sensitive layers used in the SAW sensors included in the eNose.

Samples	Solution-1 (750 µL)	Solution-2 (250 µL)
S1	ZIF-67	Au-NPs
S2	ZIF-8	Au-NPs
S3	ZIF-67	H_2O-DI
S4	ZIF-8	H_2O-DI

Figure 3. Process sequence of sensitive layer deposition on LW devices using the spin coating technique.

2.7. Experimental Setup

The sensor array was tested for acetone, ethanol, and ammonia gas analytes diluted in synthetic dry air to obtain concentrations of: 5 ppm, 10 ppm, and 25 ppm. The gas sample generator (Figure 4) consists of two mass flow controllers which were used to obtaining concentrations at a constant flow of 100 ml/min. Each array sensor works in an oscillator circuit, which includes an amplifier and a directional coupler. Therefore, when the sensor is perturbed, the oscillating frequency is shifted. A heterodyne configuration was used for signal acquisition, mixing the signal of the oscillator coupled to a reference sensor of the array (local oscillator) with the signal from an oscillator coupled to a selected ZIF nanocrystal-based SAW sensor (input signal). The frequencies obtained from the mixer were lower than 1 MHz and were acquired by a microcontroller programed as a frequency counter. For the detection, a sensing system composed of a measuring instrument (eNose) manufactured to be operated with a SAW sensor array was used. Finally, the sensor array response was acquired by a micro-frequency counter, and the information was transmitted wirelessly by a radio module to a PC with a custom application developed to display and store experimental the sensor data in real time.

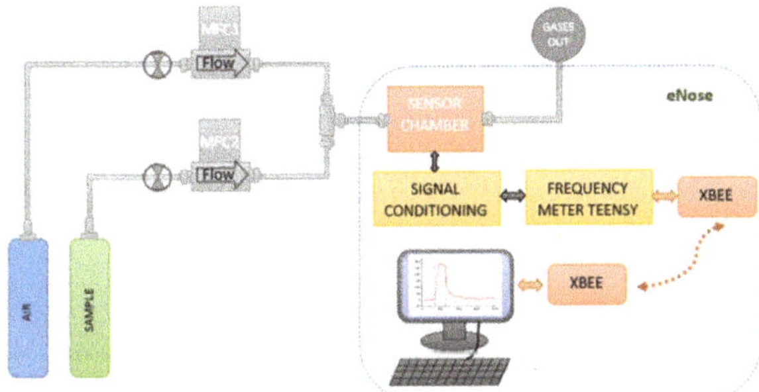

Figure 4. Experimental setup for the eNose characterization.

3. Results and Discussion

3.1. Structural and Morphological Characterization of ZIFs

The FTIR spectra of KBr diluted ZIF-8 and ZIF-67 samples show the characteristic adsorption bands of 2meImidazole ring vibrations, reported for these structures in previous works [42] (Figure 5a). The XRD patterns of both ZIFs samples show evidence for the formation of a largely crystalline structure with long-range order (Figure 5b). The position and relative intensity of the diffraction maxima are in agreement with the literature for ZIF-8 and ZIF-67 frameworks [38,39,42].

SEM images from the ZIFs layers show that the nanocrystals were configured as a continuous layer with very small particles (Figure 6a). This fact is important because nanostructured layers have two main advantages: first, the surface area of interaction with the gaseous environment is higher; and second, the SAW propagates in a continuous layer with very low scattering losses because to its wavelength (28 μm) is much larger than the diameter of the nanocrystals. The micrographs also showed hexagonal shaped nanocrystals of 50 nm approximately for the ZIF-8 samples (Figure 6b) and 200 nm approximately for the ZIF-67 samples (Figure 6c).

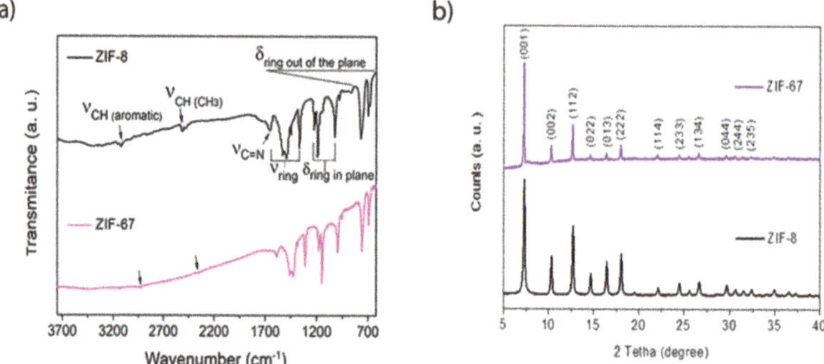

Figure 5. (**a**) FTIR spectra of the ZIF-8 and ZIF-67 samples diluted in KBr. (**b**) 2- XRD Powder Pattern of the ZIF-8 and ZIF-67 samples.

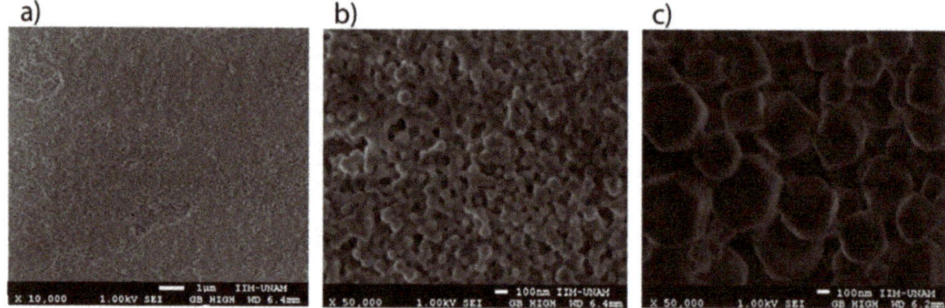

Figure 6. SEM image with magnification of (**a**) 10,000× for a continuous layer of ZIF-8 nanocrystals, (**b**) 50,000× for a layer of ZIF-8 nanocrystals and (**c**) 50,000× for a layer of ZIF-67 nanocrystals.

3.2. Electrical Characterization of the Love-Wave Sensors

The LW sensors were characterized before and after depositing the ZIF nanocrystal sensitive layers. In the array, a reference device without a sensitive layer was used (Figure 7a) to compensate sensor responses for undesirable changes in temperature and pressure. The sensors were characterized by means of the automatic network analyzer (ANA Wiltron 360B, WILTRON CO., Ltd., Incheon, Korea) and the S21 parameter was used to measure insertion loss transmission. The frequency response of each sensor exhibited a frequency shift of the minimum insertion loss caused by the mass loading of the sensitive material (Figure 7b,c). On the other hand, increases of insertion losses (Table 2) were a consequence of the scattering due to the propagation of the SAW wave in the sensitive material.

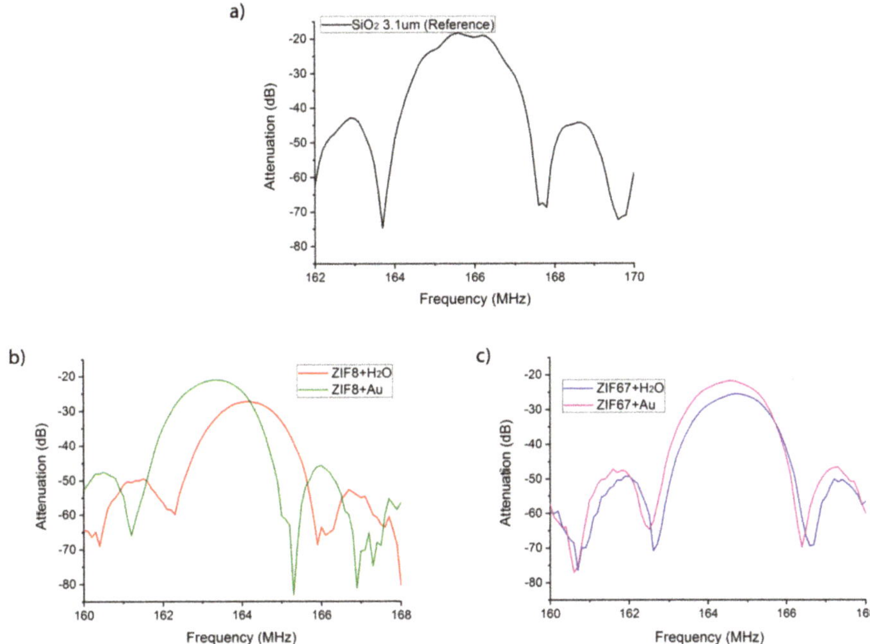

Figure 7. Spectral response of the LW sensors (**a**) without (**b**) with ZIF8 and (**c**) with ZIF67 layers.

Table 2. Insertion loss and frequency shift with respect to the reference sensor, 165 MHz and 18.3 dB.

Sensors	Sensitive Layer	Attenuation (dB)	Frequency Shift (Hz)
S1	ZIF-67 + AuNPs	3.4	1100
S2	ZIF-8 + AuNPs	2.6	2400
S3	ZIF-67 + H2O	7.1	900
S4	ZIF-8 + H2O	8.7	1500

3.3. Gas Characterization

The SAW eNose based on a LW sensor array with ZIFs nanocrystals was tested with acetone, ethanol and ammonia markers related to diabetes mellitus disease. The sensors were exposed for two minutes to each analyte at concentrations of 5 ppm, 10 ppm and 25 ppm, and then the array was purged with dry synthetic air for 10 min. The LW sensors showed a notable and fast response, e.g., for 10 ppm of acetone a frequency shift of 275 Hz and a τ_{90}, around 30 s, with a complete recovery achieved after 10 min (Figure 8a), τ_{90} being defined as the time taken to reach 90% of the frequency shift. The response of the sensor array to different concentrations of acetone (Figure 8b), ethanol (Figure 8c), and ammonia (Figure 8d) showed a high frequency shift for the different sensitive layers tested, obtaining best sensitivities for S_2 (ZIF8_Au). The measurement reproducibility was tested in two different forms; first, the lowest concentration (5 ppm) was measured twice in continuous cycles, during which a similar frequency shift was obtained; and second, the measurement with 25 ppm of ammonia was repeated three times, as a control test (Figure 8e), showing similar values of frequency shifts. Therefore, the eNose can be used for measurements in a few seconds and repetitions or new measurements can be carried out after ten minutes.

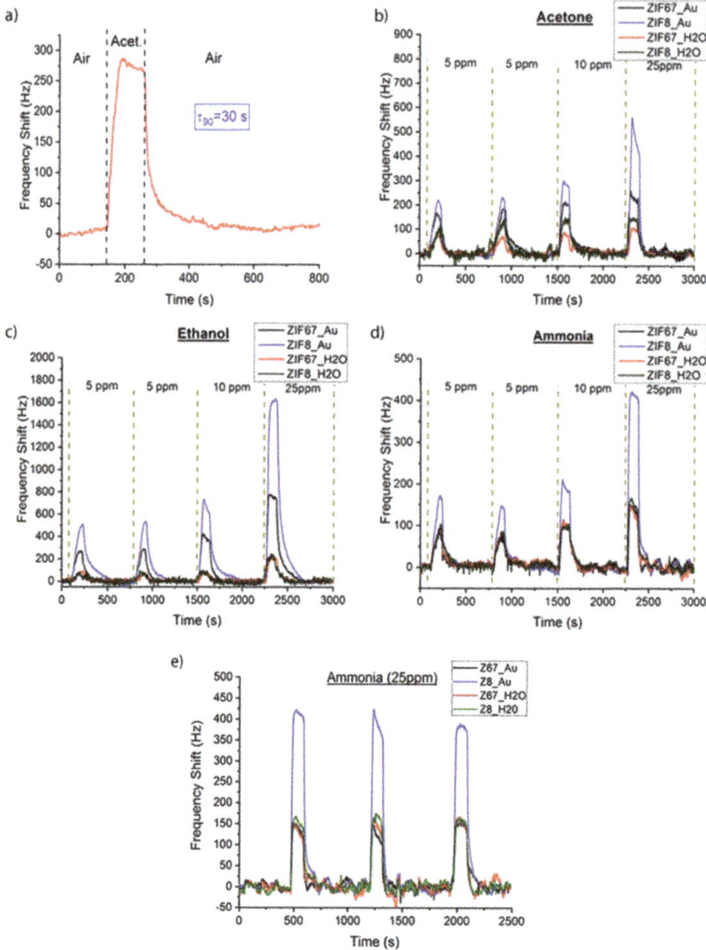

Figure 8. (**a**) SAW eNose response to 10 ppm of acetone. Experimental response for 5 ppm, 10 ppm and 25 ppm of (**b**) acetone, (**c**) ethanol and (**d**) ammonia with S1, S2, S3, and S4 sensitive layers. (**e**) Three consecutive measurements for 25 ppm of ammonia.

The reference sensor helped to compensate in the sensor array the pressure and temperature changes due to external factors, measuring only variations related to the interaction of the analytes with the sensitive layers (Table 3). The frequency shifts at the end of the exposure time were taken as responses and calibration curves were obtained (Figure 9). The responses of the sensors were found to increase with higher concentrations.

Table 3. Sensitivity, limit of detection and response time of the sensors S1, S2, S3 and S4 obtained from responses of 10 ppm.

Sensor	Sensitivity (Hz/ppm)			Limit of Detection (ppm)			Response Time (s)		
	Acet.	Etha.	Ammo.	Acet.	Etha.	Ammo.	Acet.	Etha.	Ammo.
S1	20	38	9	1.5	0.8	3.2	33	30	24
S2	28	72	19	1.1	0.5	1.6	30	40	27
S3	8	10	11	3.6	3.0	2.9	36	38	38
S4	15	11	10	2.1	2.8	2.9	44	36	41

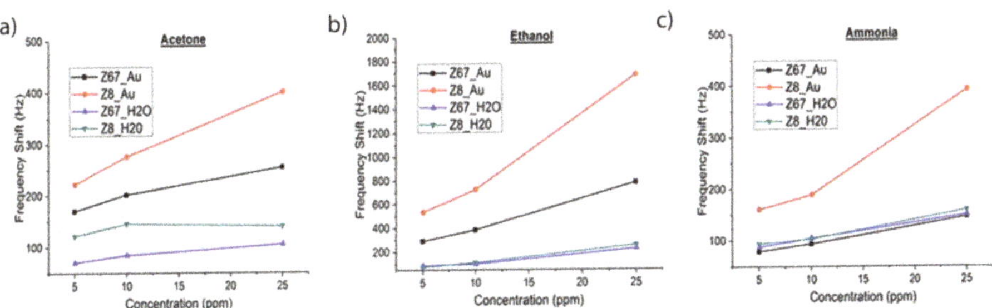

Figure 9. Calibration curves of the sensors S1, S2, S3 and S4 for (**a**) acetone, (**b**) ethanol and (**c**) ammonia.

The ZIFs/AuNP layers deposited on SiO_2 showed better mechanical properties than the ZIFs layers. In spite of a larger shift of the SAW response towards a lower frequency, they had lower insertion losses, which explains the enhanced gas-sensing properties of ZIFs/AuNP layers for the three samples. However, in the cases of acetone and ethanol, the ZIFs/AuNP layers had a significantly increased sensitivity that can be attributed to the strong interaction Au-gas molecule [43]. On the other hand, the greater sensitivity of the ZIFs/AuNP layers for ethanol can be associated with the well-known catalytic activity of <5 nm sized AuNPs for selective oxidation of alcohols [44]. Ethanol adsorbed on the AuNPs surface can then be oxidized, even at room conditions, to acetaldehyde which is detected with enhanced sensitivity by the ZIFs [21,45].

The present sensor array with nanocrystalline ZIFs as sensitive materials allows detection and discrimination of acetone, ammonia, and ethanol, as sensor responses show in the radial surface analysis to 10 ppm of the three markers (Figure 10a). In order to make discrimination and classification more feasible in a real case, for a given marker concertation the response for each sensor was normalized to the sum of the responses of the different sensors. By doing so, the information in the data of all sensors is conserved. Principal component analysis (PCA) was applied to this data and a total separation is observed among the markers (Figure 10b). This shows that the SAW eNose presented can be used to relate specific fingerprints to diabetes.

The present sensors have some significant advantages over the chemoresistive sensors based on ZIFs. First, SAW sensors show high sensitivity towards acetone, ethanol and ammonia at room temperature and chemoresistive sensors possess good sensitivity at high temperatures but poor sensitivity at room temperature [21,38,45]. Second, ZIF-8 has a high electrical resistance, requiring its combination with other materials to develop chemoresistive sensors [21,38]. Finally, a chemoresistive sensor based on ZnO nanorods showed gas selective of gases when combined with ZIF–8 shell [21], decreasing sensitivity towards acetone, ethanol or ammonia; however, SAW devices with ZIF sensitive layers combined with nanoparticles improve the sensitivity towards acetone and ethanol.

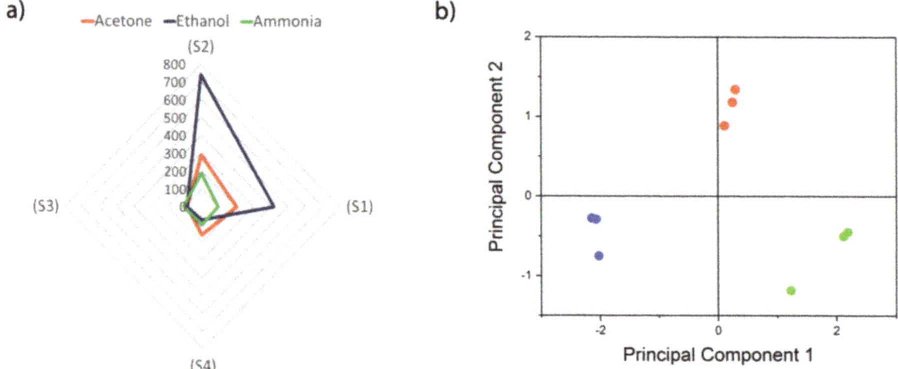

Figure 10. (**a**) Radial representation of the sensor array responses to 10 ppm of acetone, ethanol and ammonia. (**b**) Principal components analysis applied to data for discrimination of acetone (red), ethanol (blue) and ammonia (green).

4. Conclusions

A SAW eNose based on Love-wave sensors combined with ZIF-8, ZIF-67, ZIF-8/AuNP and ZIF-67/AuNP as sensitive layers was tested to three breath markers of Diabetes mellitus: acetone, ethanol and ammonia at concentrations of 5 ppm, 10 ppm and 25 ppm.

It has been shown that the SAW/ZIF eNose is effective in obtaining high sensitivity, selectivity, and reproducibility. Fast detection and recovery responses have been achieved as well, detecting concentrations as low as 5 ppm of ammonia. Finally, the efficiency of the system to discriminate Diabetes mellitus markers has been demonstrated using principal component analysis.

In conclusion, tests carried out in this work exhibited a properly performance of the SAW/ZIF eNose to be proved in future works as a prototype for a noninvasive system contributing to the diagnosis and control of Diabetes mellitus in real cases.

Author Contributions: F.A.B. and D.M. designed and developed the instrumentation, and they measured the disease markers. A.S.V. and J.M.S. designed and synthesized the nanomaterials. C.S.P., J.M.S. and D.M. designed and directed the research. I.G. fabricated the SAW devices. M.M.S.A. provided a medical overview focus on the application of the eNose for the diabetes diagnosis and control. All authors have participated in the discussions of the results and have approved the final version of the manuscript.

Funding: This work has been supported by Universidad Nacional Autónoma de México via Grants DGAPA-UNAM-PAPIIT TA100118 and DGAPA-UNAM-PAPIIT IT100518, the Fundación General CSIC via Programa ComFuturo, and the Spanish Ministry of Science and Innovation via Grant TEC2016-79898-C6-(AEI/FEDER,EU). This research has used the Spanish ICTS Network MICRONANOFABS (partially funded by MINECO).

Acknowledgments: F.A.B. thanks to CONACYT-México for his master. A.S.V. thanks to CONACYT- México by her posdoctoral grant. A.S.V. acknowledge CONACYT project 2014-Fronteras 2016-01 for the postdoctoral fellowship. F.A.B., A.S.V., J.M.S. and D.M. thank to the Laboratorio Universitario de Caracterización Espectrocópica (LUCE) by the use of its facilities and Selene Islas, V. Maturano, M. E. Mata Zamora and J.O. Flores-Flores for their technical assistance, as well as to Antonio Morales for the DRX measurements at the IF-UNAM and to Omar Novelo for the SEM and EDS measurements at the IIM-UNAM.

Conflicts of Interest: There are no conflicts of interest.

References

1. Dummer, J.; Storer, M.; Swanney, M.; McEwan, M.; Scott-Thomas, A.; Bhandari, S.; Chambers, S.; Dweik, R.; Epton, M. Analysis of biogenic volatile organic compounds in human health and disease. *TrAC Trends Anal. Chem.* **2011**, *30*, 960–967. [CrossRef]
2. Cikach, F.S.; Dweik, R.A. Cardiovascular Biomarkers in Exhaled Breath. *Prog. Cardiovasc. Dis.* **2012**, *55*, 34–43. [CrossRef] [PubMed]

3.   Miekisch, W.; Schubert, J.K. From highly sophisticated analytical techniques to life-saving diagnostics: Technical developments in breath analysis. *TrAC Trends Anal. Chem.* **2006**, *25*, 665–673. [CrossRef]

4.   Das, S.; Pal, S.; Mitra, M. Significance of Exhaled Breath Test in Clinical Diagnosis: A Special Focus on the Detection of Diabetes Mellitus. *J. Med. Biol. Eng.* **2016**, *36*. [CrossRef] [PubMed]

5.   Diskin, A.M.; Španěl, P.; Smith, D. Time variation of ammonia, acetone, isoprene and ethanol in breath: A quantitative SIFT-MS study over 30 days. *Physiol. Meas.* **2003**, *24*, 107–119. [CrossRef] [PubMed]

6.   Rheaume, J.M.; Pisano, A.P. A review of recent progress in sensing of gas concentration by impedance change. *Ionics* **2011**, *17*, 99–108. [CrossRef]

7.   Santos, J.P.; Fernández, M.J.; Fontecha, J.L.; Matatagui, D.; Matatagui, D.; Sayago, I.; Horrillo, M.C.; Gracia, I. Nanocrystalline tin oxide nanofibers deposited by a novel focused electrospinning method. Application to the detection of TATP precursors. *Sensors* **2014**, *14*, 24231–24243. [CrossRef]

8.   Nayak, A.K.; Ghosh, R.; Santra, S.; Guha, P.K.; Pradhan, D. Hierarchical nanostructured WO_3–SnO_2 for selective sensing of volatile organic compounds. *Nanoscale* **2015**, *7*, 12460–12473. [CrossRef]

9.   Hodgkinson, J.; Tatam, R.P. Optical gas sensing: A review. *Meas. Sci. Technol.* **2013**, *24*, 012004. [CrossRef]

10.   Hromadka, J.; Tokay, B.; Correia, R.; Morgan, S.P.; Korposh, S. Highly sensitive volatile organic compounds vapour measurements using a long period grating optical fibre sensor coated with metal organic framework ZIF-8. *Sens. Actuators B Chem.* **2018**, *260*, 685–692. [CrossRef]

11.   Fragoso-Mora, J.R.; Matatagui, D.; Bahos, F.A.; Fontecha, J.; Fernandez, M.J.; Santos, J.P.; Sayago, I.; Gràcia, I.; Horrillo, M.C. Gas sensors based on elasticity changes of nanoparticle layers. *Sens. Actuators B Chem.* **2018**, *268*, 93–99. [CrossRef]

12.   Ballantine, D.S.; Martin, S.J.; Ricco, A.J.; Frye, G.C.; Wohltjen, H.; White, R.M.; Zellers, E.T. Chapter 3-Acoustic Wave Sensors and Responses. In *Applications of Modern Acoustics*; Ballantine, D.S., Martin, S.J., Ricco, A.J., Frye, G.C., Wohltjen, H., White, R.M., Zellers, E.T., Eds.; Academic Press: Burlington, NJ, USA, 1997; pp. 36–149. ISBN 978-0-12-077460-9.

13.   Kannan, P.K.; Late, D.J.; Morgan, H.; Rout, C.S. Recent developments in 2D layered inorganic nanomaterials for sensing. *Nanoscale* **2015**, *7*, 13293–13312. [CrossRef] [PubMed]

14.   Jiménez-Cadena, G.; Riu, J.; Rius, F.X. Gas sensors based on nanostructured materials. *Analyst* **2007**, *132*, 1083–1099. [CrossRef] [PubMed]

15.   Comini, E. Metal oxide nanowire chemical sensors: Innovation and quality of life. *Mater. Today* **2016**, *19*, 559–567. [CrossRef]

16.   Xu, X.; Wang, J.; Long, Y.; Xu, X.; Wang, J.; Long, Y. Zeolite-based Materials for Gas Sensors. *Sensors* **2006**, *6*, 1751–1764. [CrossRef]

17.   Kreno, L.E.; Leong, K.; Farha, O.K.; Allendorf, M.; Van Duyne, R.P.; Hupp, J.T. Metal–Organic Framework Materials as Chemical Sensors. *Chem. Rev.* **2012**, *112*, 1105–1125. [CrossRef] [PubMed]

18.   DMello, M.E.; Sundaram, N.G.; Kalidindi, S.B. Assembly of ZIF-67 Metal-Organic Framework over Tin Oxide Nanoparticles for Synergistic Chemiresistive CO_2 Gas Sensing. *Chem. A Eur. J.* **2018**, *24*, 9220–9223. [CrossRef]

19.   Paschke, B.; Wixforth, A.; Denysenko, D.; Volkmer, D. Fast Surface Acoustic Wave-Based Sensors to Investigate the Kinetics of Gas Uptake in Ultra-Microporous Frameworks. *ACS Sens.* **2017**, *2*, 740–747. [CrossRef]

20.   Devkota, J.; Kim, K.-J.; Ohodnicki, P.R.; Culp, J.T.; Greve, D.W.; Lekse, J.W. Zeolitic imidazolate framework-coated acoustic sensors for room temperature detection of carbon dioxide and methane. *Nanoscale* **2018**, *10*, 8075–8087. [CrossRef]

21.   Tian, H.; Fan, H.; Li, M.; Ma, L. Zeolitic Imidazolate Framework Coated ZnO Nanorods as Molecular Sieving to Improve Selectivity of Formaldehyde Gas Sensor. *ACS Sens.* **2016**, *1*, 243–250. [CrossRef]

22.   Zhang, K.; Lively, R.P.; Zhang, C.; Chance, R.R.; Koros, W.J.; Sholl, D.S.; Nair, S. Exploring the Framework Hydrophobicity and Flexibility of ZIF-8: From Biofuel Recovery to Hydrocarbon Separations. *J. Phys. Chem. Lett.* **2013**, *4*, 3618–3622. [CrossRef]

23.   Eum, K.; Jayachandrababu, K.C.; Rashidi, F.; Zhang, K.; Leisen, J.; Graham, S.; Lively, R.P.; Chance, R.R.; Sholl, D.S.; Jones, C.W.; et al. Highly Tunable Molecular Sieving and Adsorption Properties of Mixed-Linker Zeolitic Imidazolate Frameworks. *J. Am. Chem. Soc.* **2015**, *137*, 4191–4197. [CrossRef]

24. Tao, J.; Wang, X.; Sun, T.; Cai, H.; Wang, Y.; Lin, T.; Fu, D.; Ting, L.L.Y.; Gu, Y.; Zhao, D. Hybrid Photonic Cavity with Metal-Organic Framework Coatings for the Ultra-Sensitive Detection of Volatile Organic Compounds with High Immunity to Humidity. *Sci. Rep.* **2017**, *7*. [CrossRef] [PubMed]
25. Mayer-Davis, E.J.; Lawrence, J.M.; Dabelea, D.; Divers, J.; Isom, S.; Dolan, L.; Imperatore, G.; Linder, B.; Marcovina, S.; Pettitt, D.J.; et al. Incidence Trends of Type 1 and Type 2 Diabetes among Youths, 2002–2012. *N. Engl. J. Med.* **2017**, *376*, 1419–1429. [CrossRef] [PubMed]
26. Letourneau, L.R.; Carmody, D.; Wroblewski, K.; Denson, A.M.; Sanyoura, M.; Naylor, R.N.; Philipson, L.H.; Greeley, S.A.W. Diabetes Presentation in Infancy: High Risk of Diabetic Ketoacidosis. *Diabetes Care* **2017**, *40*, e147–e148. [CrossRef] [PubMed]
27. Anderson, J.C.; Lamm, W.J.E.; Hlastala, M.P. Measuring airway exchange of endogenous acetone using a single-exhalation breathing maneuver. *J. Appl. Physiol.* **2006**, *100*, 880–889. [CrossRef] [PubMed]
28. Freund, G.; Weinsier, R.L. Standardized ketosis in man following medium chain triglyceride ingestion. *Metabolism* **1966**, *15*, 980–991. [CrossRef]
29. Saslow, L.R.; Kim, S.; Daubenmier, J.J.; Moskowitz, J.T.; Phinney, S.D.; Goldman, V.; Murphy, E.J.; Cox, R.M.; Moran, P.; Hecht, F.M. A Randomized Pilot Trial of a Moderate Carbohydrate Diet Compared to a Very Low Carbohydrate Diet in Overweight or Obese Individuals with Type 2 Diabetes Mellitus or Prediabetes. *PLoS ONE* **2014**, *9*. [CrossRef]
30. Phinney, S.D.; Bistrian, B.R.; Wolfe, R.R.; Blackburn, G.L. The human metabolic response to chronic ketosis without caloric restriction: Physical and biochemical adaptation. *Metabolism* **1983**, *32*, 757–768. [CrossRef]
31. Sulway, M.J.; Malins, J.M. ACETONE IN DIABETIC KETOACIDOSIS. *Lancet* **1970**, *296*, 736–740. [CrossRef]
32. Minh, T.D.C.; Oliver, S.R.; Ngo, J.; Flores, R.; Midyett, J.; Meinardi, S.; Carlson, M.K.; Rowland, F.S.; Blake, D.R.; Galassetti, P.R. Noninvasive measurement of plasma glucose from exhaled breath in healthy and type 1 diabetic subjects. *Am. J. Physiol. Endocrinol. Metab.* **2011**, *300*, E1166–E1175. [CrossRef] [PubMed]
33. Galassetti, P.R.; Novak, B.; Nemet, D.; Rose-Gottron, C.; Cooper, D.M.; Meinardi, S.; Newcomb, R.; Zaldivar, F.; Blake, D.R. Breath Ethanol and Acetone as Indicators of Serum Glucose Levels: An Initial Report. *Diabetes Technol. Ther.* **2005**, *7*, 115–123. [CrossRef] [PubMed]
34. Lee, J.; Ngo, J.; Blake, D.; Meinardi, S.; Pontello, A.M.; Newcomb, R.; Galassetti, P.R. Improved predictive models for plasma glucose estimation from multi-linear regression analysis of exhaled volatile organic compounds. *J. Appl. Physiol.* **2009**, *107*, 155–160. [CrossRef] [PubMed]
35. Qin, T.; Xu, X.; Polák, T.; Pacáková, V.; Štulík, K.; Jech, L. A simple method for the trace determination of methanol, ethanol, acetone and pentane in human breath and in the ambient air by preconcentration on solid sorbents followed by gas chromatography. *Talanta* **1997**, *44*, 1683–1690. [CrossRef]
36. Turner, C.; Španěl, P.; Smith, D. A longitudinal study of ammonia, acetone and propanol in the exhaled breath of 30 subjects using selected ion flow tube mass spectrometry, SIFT-MS. *Physiol. Meas.* **2006**, *27*, 321–337. [CrossRef] [PubMed]
37. Schmidt, F.M.; Vaittinen, O.; Metsälä, M.; Lehto, M.; Forsblom, C.; Groop, P.-H.; Halonen, L. Ammonia in breath and emitted from skin. *J. Breath Res.* **2013**, *7*. [CrossRef] [PubMed]
38. Matatagui, D.; Sainz-Vidal, A.; Gràcia, I.; Figueras, E.; Cané, C.; Saniger, J.M. Chemoresistive gas sensor based on ZIF-8/ZIF-67 nanocrystals. *Sens. Actuators B Chem.* **2018**, *274*, 601–608. [CrossRef]
39. Pan, Y.; Liu, Y.; Zeng, G.; Zhao, L.; Lai, Z. Rapid synthesis of zeolitic imidazolate framework-8 (ZIF-8) nanocrystals in an aqueous system. *Chem. Commun.* **2011**, *47*, 2071–2073. [CrossRef]
40. Gross, A.F.; Sherman, E.; Vajo, J.J. Aqueous room temperature synthesis of cobalt and zinc sodalite zeolitic imidizolate frameworks. *Dalt. Trans.* **2012**, *41*, 5458–5460. [CrossRef]
41. Sahu, S.R.; Devi, M.M.; Mukherjee, P.; Sen, P.; Biswas, K. Optical Property Characterization of Novel Graphene-X (X = Ag, Au and Cu) Nanoparticle Hybrids. *J. Nanomater.* **2013**, *2013*, 1–9. [CrossRef]
42. Park, K.S.; Ni, Z.; Côté, A.P.; Choi, J.Y.; Huang, R.; Uribe-Romo, F.J.; Chae, H.K.; O'Keeffe, M.; Yaghi, O.M. Exceptional chemical and thermal stability of zeolitic imidazolate frameworks. *Proc. Natl. Acad. Sci. USA* **2006**, *103*, 10186–10191. [CrossRef] [PubMed]
43. Xia, J.; Diao, K.; Zheng, Z.; Cui, X. Porous Au/ZnO nanoparticles synthesised through a metal organic framework (MOF) route for enhanced acetone gas-sensing. *RSC Adv.* **2017**, *7*, 38444–38451. [CrossRef]

44. Sharma, A.S.; Kaur, H.; Shah, D. Selective oxidation of alcohols by supported gold nanoparticles: Recent advances. *RSC Adv.* **2016**, *6*, 28688–28727. [CrossRef]

45. Chen, E.-X.; Yang, H.; Zhang, J. Zeolitic Imidazolate Framework as Formaldehyde Gas Sensor. *Inorg. Chem.* **2014**, *53*, 5411–5413. [CrossRef] [PubMed]

 *biosensors* 

Article

Non-Invasive Diagnosis of Diabetes by Volatile Organic Compounds in Urine Using FAIMS and Fox4000 Electronic Nose

Siavash Esfahani [1,*], Alfian Wicaksono [1], Ella Mozdiak [2], Ramesh P. Arasaradnam [2,3,4] and James A. Covington [1]

1 School of Engineering, University of Warwick, Coventry CV4 7AL, UK; a.wicaksono@warwick.ac.uk (A.W.); J.A.Covington@warwick.ac.uk (J.A.C.)
2 Department of Gastroenterology, University Hospital Coventry and Warwickshire, Coventry, CV2 2DX, UK; ella.mozdiak@nhs.net (E.M.); R.Arasaradnam@warwick.ac.uk (R.P.A.)
3 School of Applied Biological Sciences, University of Coventry, Coventry CV1 5FB, UK
4 Warwick Medical School, University of Warwick, Coventry CV4 7AL, UK
* Correspondence: S.Esfahani@warwick.ac.uk or siavash_esfahani6104@yahoo.com; Tel.: +44-750-222-1241

Received: 9 October 2018; Accepted: 19 November 2018; Published: 1 December 2018

Abstract: The electronic nose (eNose) is an instrument designed to mimic the human olfactory system. Usage of eNose in medical applications is more popular than ever, due to its low costs and non-invasive nature. The eNose sniffs the gases and vapours that emanate from human waste (urine, breath, and stool) for the diagnosis of variety of diseases. Diabetes mellitus type 2 (DM2) affects 8.3% of adults in the world, with 43% being underdiagnosed, resulting in 4.9 million deaths per year. In this study, we investigated the potential of urinary volatile organic compounds (VOCs) as novel non-invasive diagnostic biomarker for diabetes. In addition, we investigated the influence of sample age on the diagnostic accuracy of urinary VOCs. We analysed 140 urine samples (73 DM2, 67 healthy) with Field-Asymmetric Ion Mobility Spectrometry (FAIMS); a type of eNose; and FOX 4000 (AlphaM.O.S, Toulouse, France). Urine samples were collected at UHCW NHS Trust clinics over 4 years and stored at −80 °C within two hours of collection. Four different classifiers were used for classification, specifically Sparse Logistic Regression, Random Forest, Gaussian Process, and Support Vector on both FAIMS and FOX4000. Both eNoses showed their capability of diagnosing DM2 from controls and the effect of sample age on the discrimination. FAIMS samples were analysed for all samples aged 0–4 years (AUC: 88%, sensitivity: 87%, specificity: 82%) and then sub group samples aged less than a year (AUC (Area Under the Curve): 94%, Sensitivity: 92%, specificity: 100%). FOX4000 samples were analysed for all samples aged 0–4 years (AUC: 85%, sensitivity: 77%, specificity: 85%) and a sub group samples aged less than 18 months: (AUC: 94%, sensitivity: 90%, specificity: 89%). We demonstrated that FAIMS and FOX 4000 eNoses can discriminate DM2 from controls using urinary VOCs. In addition, we showed that urine sample age affects discriminative accuracy.

Keywords: electronic nose; biosensor; diabetes; FOX 4000; FAIMS; urine sample; non-invasive diagnosis; medical application; volatile organic compounds (VOCs)

---

1. Introduction

The growing rate of diabetes and its related diseases is becoming a worldwide major health concern. The motivation of this paper was to make use of a technology called the "electronic nose" (eNose) for diagnosing diabetes. Using eNose technology with urinary volatile organic compounds (VOCs) is attractive as it allows non-invasive monitoring of various molecular constituents in urine. Trace gases in urine are linked to metabolic reactions and diseases.

The mimicry of a biological olfactory system, called the electronic nose, was developed in the early 1980s [1]. The electronic nose contains arrays of sensors that analyses the sample as a whole complex mixture, not identifying a specific chemical [2,3]. By developing technology and increasing demand for non-invasive methods for diagnosis diseases, the electronic nose is becoming a promising instrument in the medical domain. Commercial and experimental electronic noses have been developed for diagnosis of a wide range of diseases such as lung cancer [4], breast cancer [5], brain cancer [6] and melanoma [7], prostate cancer [8], colorectal cancer [2], asthma [9], and many other diseases. There are only few studies on diagnosing diabetes using urinary VOCs with eNose instruments [10,11].

Currently, one of the urgent public medical issues is the fast-growing number of people with diabetes. According to statistics, the number of people worldwide with diabetes in 2017 was estimated to be 425 million, with 1 of 2 adults remaining undiagnosed [12]. In the UK, the number of people diagnosed with type 2 diabetes was just under 3.7 million people in 2017, with a further estimated 1 million people remaining undiagnosed, which is better than worldwide figures [13]. It is a major health concern especially in the under 20 s, where the numbers of diabetic children are rapidly increasing. In the UK alone, around 31,500 patients under the age of 19 have diabetes [12]. According to the National Diabetes Audit (NDA) report, 24,000 patients suffering from diabetes have early death each year (65 patients a day) [14]. From a financial cost point of view, 10% of the NHS budget is spent on diabetes.

Our approach was to undertake a pilot study to investigate if urinary VOCs (volatile organic compounds) could be used as a non-invasive means to identify patients with type 2 diabetes Mellitus (T2DM). These samples were collected over a four-and-a-half-year period and stored at $-80\,^{\circ}$C and then analysed using by Owlstone Lonestar FAIMS and FOX4000, as two types of electronic nose. From our previous study, it was discovered that samples over 12 months old will not emit sufficient VOCs for diagnostic purposes [15], hence this paper will focus more on analysing samples less than 12 months old for diagnosing diabetes samples compared to healthy control.

2. Materials and Methods

2.1. Sample Preparation

One hundred and thirty-eight patients were recruited at the University Hospital Coventry & Warwickshire, UK. Each recruit provided a urine sample, which was collected in a clinic and frozen at $-80\,^{\circ}$C within two hours over a four-and-a-half-year period. Seventy-one samples came from patients with type 2 diabetes, with a further 67 samples from healthy controls. Scientific and ethical approval was obtained from the Warwickshire Research & Development Department and Warwickshire Ethics Committee 09/H1211/38. Written informed consent was obtained from all patients who participated in the study. For analysis, samples were thawed to 4 °C in a laboratory fridge for 24 h prior to testing to minimise chemical loss. Demographic details of patients are shown in Table 1.

Table 1. Demographic information of used urinary samples (incomplete data for 2 diabetic patients).

Demographic Data	Diabetes	Control
Male (%)	27 (39.1)	43 (64.2)
Female (%)	42 (60.9)	24 (35.8)
Median age (year)	57	53.5
Mean alcohol (units/week)	1.8	1.09
Median BMI	39.7	26.1

2.2. FAIMS Chemical Analyser

A commercial FAIMS (Field-Asymmetric Ion Mobility Spectrometer) device was used in this study, specifically a Lonestar instrument (Owlstone, Cambridge, UK). It is able to separate complex chemical mixtures by measuring the difference in mobility of ionised molecules in high electric fields, thus it measures a physical property of a gas or vapour. The Lonestar was setup to use dynamic headspace sampling, using an ATLAS sampling system (Owlstone, Cambridge, UK), which controls of the flow rate and the temperature of the sample. The unit pushes clean/dry air over the surface of the urine and into the Lonestar instrument. The chemical components are then ionised (Ni-63 source) and pushed through two parallel plates. An asynchronous waveform is applied to these plates, consisting of a high electric field for a short period of time, followed by an inverse potential of low electric field, but with the time x electric field being equal. If a molecule's mobility is constant with electric field, the ion exits the plates and is detected. However, if the electric field attracts or repels an ion, it drifts towards a plate and loses its change when it makes contact. To remove this drift, a constant voltage (called the compensation voltage) is applied, thus by scanning through different compensation voltages, we can measure a range of mobilities. Both the magnitude of the electric field (called the dispersion field) and the compensation voltage is scanned to create a 3D map of molecular mobilities [16]. Figure 1 shows the FAIMS instrument setup and Figure 2 shows the typical output of FAIMS instrument. In this experiment, 5 mL of urine were aliquoted from each sample into a 10 mL glass vial and placed into an ATLAS sample system and followed a similar setup to one previously used by our group [17,18]. This heated the sample to 40 ± 0.1 °C. Each sample was tested three times sequentially, with each run having a flow rate over the sample of 200 mL/min of clean dry air. Further make-up air was added to create a total flow rate of 2 L/min. The FAIMS was scanned from 0% to 99% dispersion field in 51 steps, -6 V to +6 V compensation voltage in 512 steps and both positive and negative ions were detected to create a test file composed of 52,224 data points.

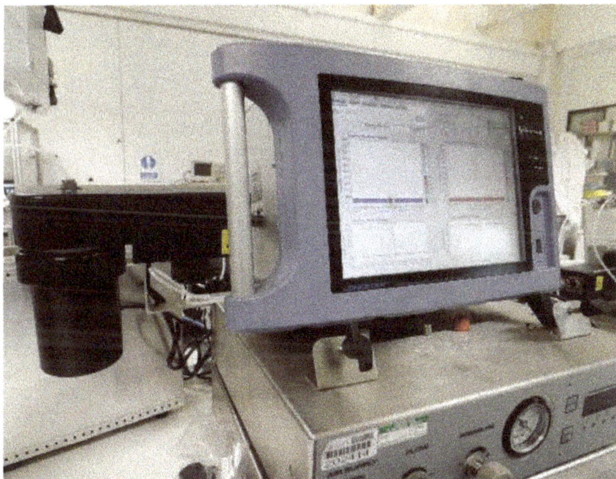

Figure 1. Field-asymmetric ion mobility spectrometer (FAIMS) setup.

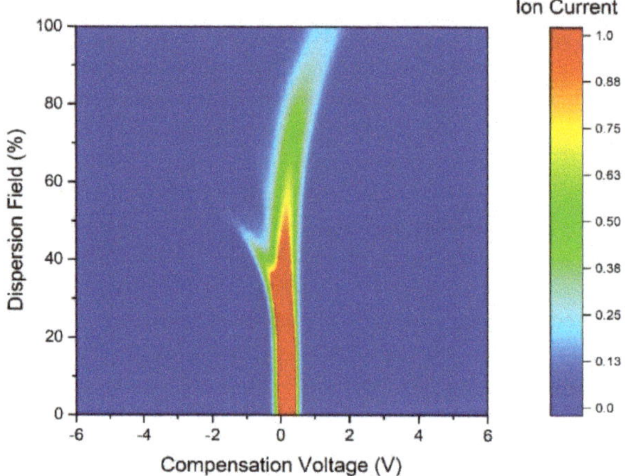

Figure 2. Typical FAIMS output responding to urine vapour.

2.3. Electronic Nose

A commercial electronic nose (FOX 4000 with HS100 autosampler, Alpha M.O.S, Toulouse, France) was used in this study. The Fox 4000 consists of an injection system, sensor chambers, mass flow controller, and acquisition board with microcontroller. The electronic nose contains 18 metal oxide gas sensors that are placed in three chambers and were calibrated regularly in line with the manufacturer's recommended procedures to ensure stability. These three chambers are called T, P, and LY. All the sensors' names and their application are indicated in Table 2.

Table 2. α-FOX4000 eNose sensor arrays and their applications.

Sensor No.	References	Description
S1	LY2/LG	Oxidising gas
S2	LY2/G	Ammonia, carbon monoxide
S3	LY2/AA	Ethanol
S4	LY2/GH	Ammonia/ Organic amines
S5	LY2/gCTL	Hydrogen sulfide
S6	LY2/gCT	Propane/Butane
S7	T30/1	Organic solvents
S8	P10/1	Hydrocarbons
S9	P10/2	Methane
S10	P40/1	Fluorine
S11	T70/2	Aromatic compounds
S12	PA/2	Ethanol, Ammonia/Organic amines
S13	P30/1	Polar compounds (Ethanol)
S14	P40/2	Heteroatom/Chloride/Aldehydes
S15	P30/2	Alcohol
S16	T40/2	Aldehydes
S17	T40/1	Chlorinated compounds
S18	TA/2	Air quality

The basic operation principle of the electronic nose is based on the sensors' electronic resistance changes in response to the presence of volatile compounds. In our case, the output response was calculated by the formula in Equation (1) [19].

$$R = (R_0 - R_T)/R_0 \tag{1}$$

where R is response of sensor, R_0 is initial resistance of metal oxide sensor at time 0, and R_T is sensor's conductance value.

Figure 3 shows the setup of the FOX 4000 instrument. Figure 4 shows a typical response of a FOX 4000 with 18 sensors to a diabetic urine sample's volatile compounds. Each curve signifies one sensor's response. The concentration and nature of the sensed molecules plus the type of metal oxide sensors used in the eNose are the three main reasons for the size of the sensor's response [9].

Figure 3. FOX 4000 setup.

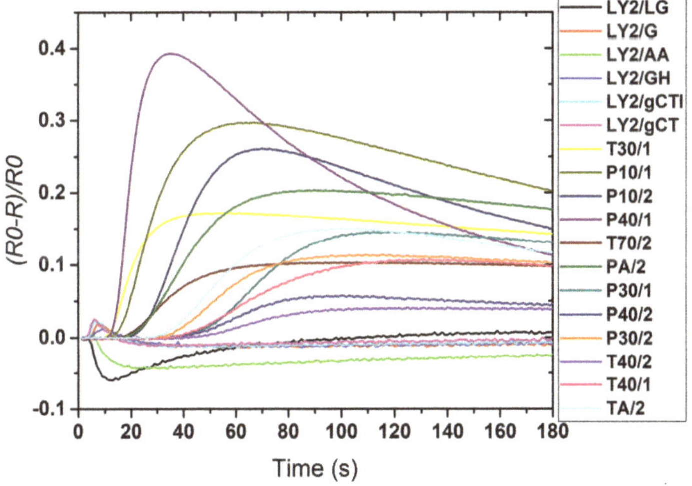

Figure 4. Typical response of a FOX 4000 to a urine sample.

The samples were agitated and heated to 40 °C for 10 min before 2.5 ml of the sample headspace was injected into the electronic nose (flow rate over the sensors was 200 ml/min of zero air, data was recorded for 180 s at a sample rate of 1 Hz). The data from the FOX4000 are generated by sampling each of the 18 metal oxide sensors at 180 data points over 180 s. These readings are concatenated into a single vector representing a single sample, and thus the raw data are of dimension 320.

3. Results

3.1. FAIMS Analysis

Each FAIMS dataset consisted of 52,224 data points that are stored in a 512×102 matrix. The first step of data processing was performing a pre-processing step by applying 2D wavelet transform (using Daubechies D4 wavelets) to each data set. This step aimed to decompose the signal and extract subtle chemical signals within a wider range of the signal. The 2D wavelet transform will concentrate the chemical information into several levels, which consist of a small number of wavelet coefficients. These coefficients would then be the input. We randomly divided the input into two sets, with 70% used as a training/validation set, and 30% as a test set. Ten-fold cross-validation was applied to the training and validation set in which, within each fold, supervised features selection was performed using Wilcoxon rank sum test by calculating the p-values for every pair of features in the training set. Principal component analysis (PCA) was performed to see distribution of data in the scatter plot. The ten most statistically important features, which had the lowest p-values, were then used to train the classifier algorithms. Four different classifiers were used for prediction, specifically Sparse Logistic Regression, Random Forest, Gaussian Process, and Support Vector Machines. The hyperparameter of the classifier was then tuned by comparing the error function using an independent validation set. This step was used to minimise overfitting. Finally, the performance of classifier algorithms and their diagnostic capabilities were calculated using an independent test set and were displayed in a graphical plot called Receiver Operator Characteristic (ROC) curve. The ROC provides information about Area Under Curve (AUC), sensitivity, specificity, Positive Predictive Value (PPV), Negative Predictive Value (NPV), and p-values. Sample age affected the vapour emission from the sample. Figure 5 shows that using the PCA method on the whole group of samples from zero to four years is not sufficient to distinguish diabetic samples from control ones. However, by separating samples by age and applying the PCA method, the results showed better separation between diabetes and control groups of newer samples. Figure 6 is related to samples of age less than 1 year, as expected, where the separation is clear between two different groups. Only three samples have cross selectivity between diabetes and control groups.

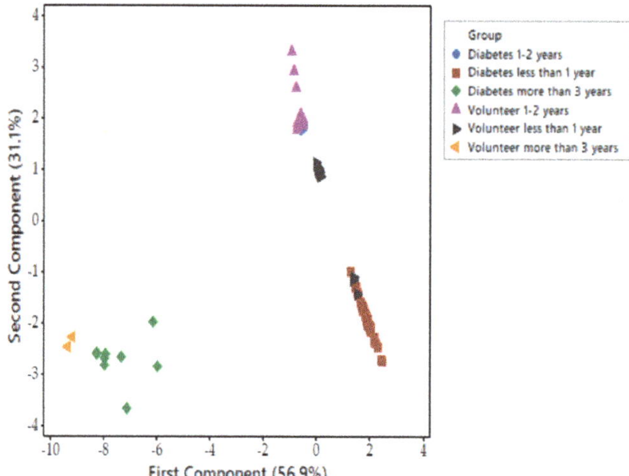

Figure 5. Principal component analysis (PCA) of samples between 0 and 4 years.

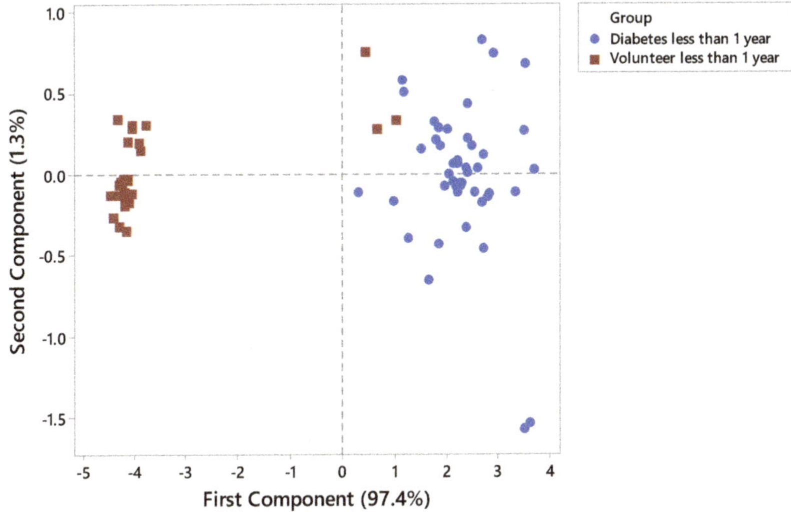

Figure 6. PCA analysis of samples less than 1 year old.

The result of ROC analysis with four different methods for samples aged 0–4 years and samples aged less than a year are summarised in Tables 3 and 4. The ROC analysis was processed for samples with four different analysis methods. Gaussian processing appears to be the best method since it has the highest area under curve and from a medical perspective, it is important to have low negative predictive value (NPV), which this method has compared to the others. The area under the curve in the ROC shows how good the separation is. The max value for area under the curve in the ROC is 1. AUC values for all methods, for data with less than a year, is more than 0.9, and it is below 0.9 for samples aged 0–4 years. Figures 7 and 8 show the ROC analysis of the two groups of sample age. As is clear from the figures, better classification performance is achieved for samples with less storage time.

Table 3. Summary of the ROC (Receiver Operator Characteristic) analysis details for samples 0–4 years old.

Methods	AUC	Sensitivity	Specificity	PPV	NPV	*p*-Value
Sparse Logistic Regression	0.89 (0.79–0.99)	0.74 (0.51–0.9)	0.88 (0.63–0.99)	0.89	0.71	4.368×10^{-6}
Random Forest	0.86 (0.74–0.98)	0.78 (0.56–0.92)	0.82 (0.56–0.96)	0.86	0.74	6.690×10^{-5}
Gaussian Process	0.88 (0.76–1)	0.87 (0.66–0.97)	0.82 (0.56–0.96)	0.87	0.82	7.187×10^{-6}
Support Vector Machine	0.88 (0.77–0.99)	0.74 (0.51–0.9)	0.94 (0.71–0.99)	0.94	0.73	7.189×10^{-6}

Table 4. Summary of the ROC analysis details for samples less than 1 year old.

Methods	AUC	Sensitivity	Specificity	PPV	NPV	*p*-Value
Sparse Logistic Regression	0.9 (0.7–1)	1 (0.75–1)	0.9 (0.55–0.99)	0.93	1	3.199×10^{-4}
Random Forest	0.93 (0.79–1)	1 (0.75–1)	0.9 (0.55–0.98)	0.93	1	1.419×10^{-4}
Gaussian Process	0.94 (0.82–1)	0.92 (0.64–1)	1 (0.69–1)	1	0.91	5.856×10^{-5}
Support Vector Machine	0.9 (0.7–1)	1 (0.75–1)	0.9 (0.55–0.99)	0.93	1	3.199×10^{-4}

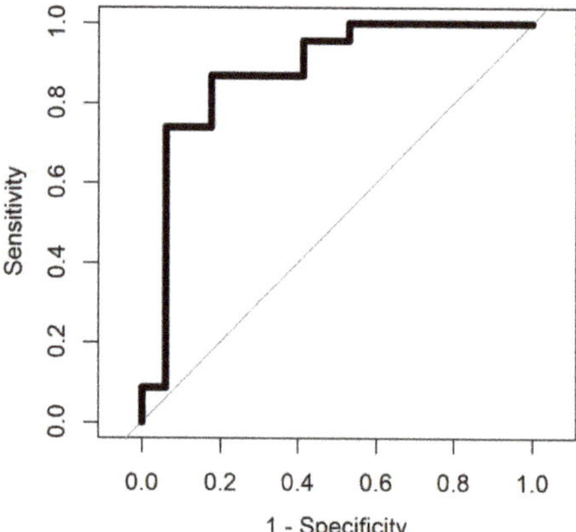

Figure 7. Receiver operating characteristic (ROC) analysis of samples aged 0–4 years. Gaussian process (AUC = 0.88).

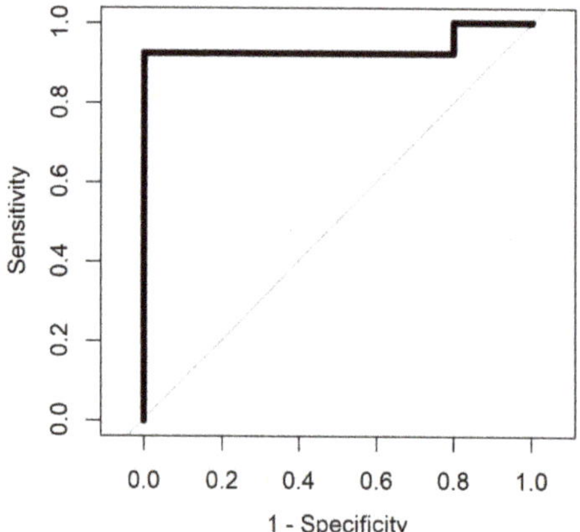

Figure 8. Receiver operating characteristic (ROC) analysis of samples aged less than a year. Gaussian Process (AUC = 0.94).

3.2. Electronic Nose Analysis

PCA analysis was used for classification of the features that were extracted by dividing the maximum resistance by the baseline resistance. Plotting only the first and second PCA components shows the disease classification has been affected by the urine sample's storage age. Figure 9 illustrates the diabetes and control samples collected 4 years prior to analysis, for which the classification is not performed appropriately. Figure 10 shows the classification for diabetes and control samples collected and tested in less than 18 months. The results show that newer samples are tightly clustered and sufficiently separated from the disease class in comparison to the group of older samples.

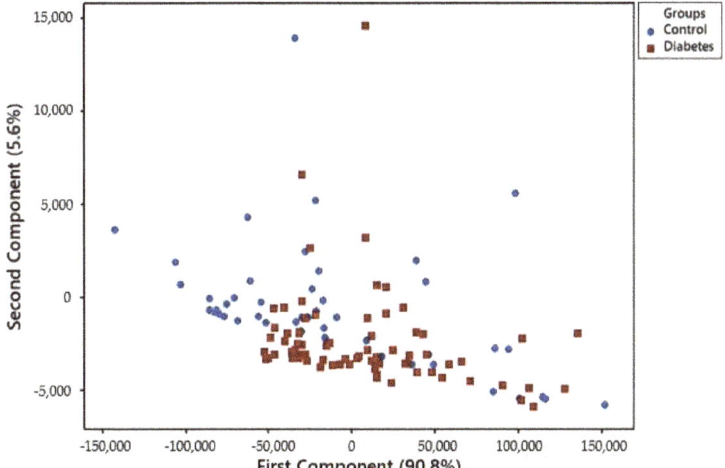

Figure 9. PCA components 1 and 2 show 4-year-old disease and control samples separation.

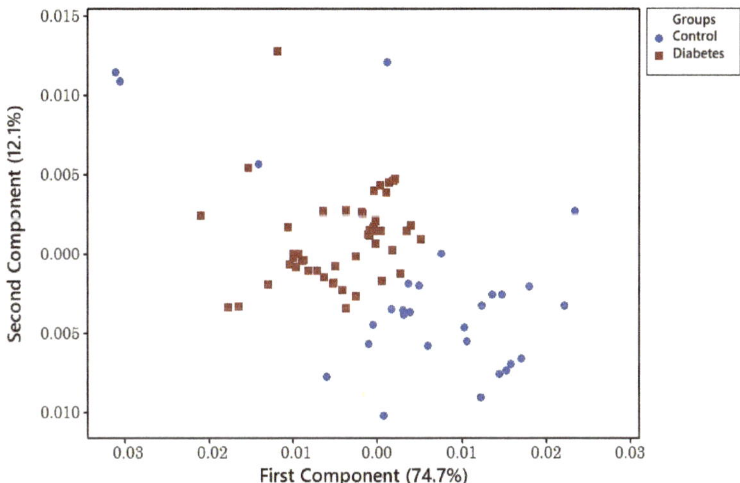

Figure 10. PCA components 1 and 2 show 18-month-old disease and control samples separation.

Figure 11 shows the linear discrimination analysis (LDA) result for the four-year-old samples. As can be seen, there is no clear separation between groups. Figure 12 shows the LDA method's result for newer samples with less than 18 months of age. This shows clear separation between both groups.

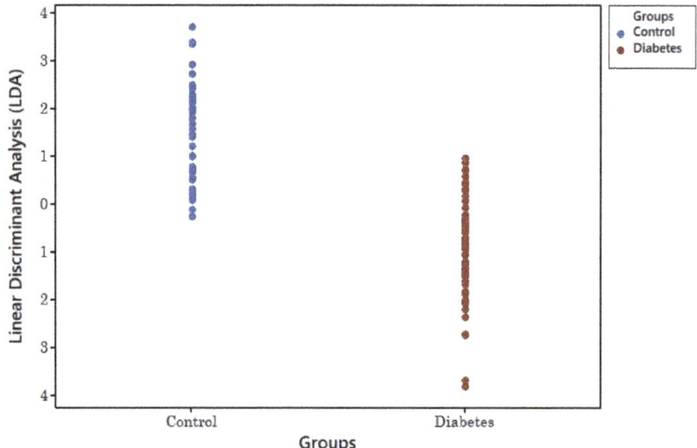

Figure 11. LDA classification of 4-year-old disease and control samples.

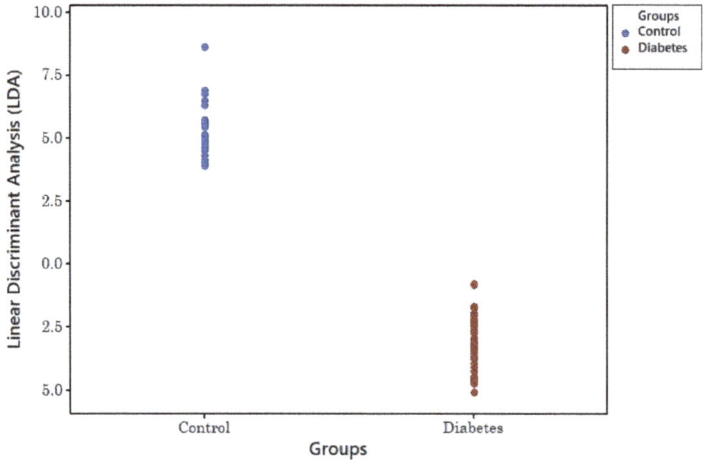

Figure 12. LDA classification of 18-month-old disease and control samples.

To analyse this set of data, maximum variances of sensor resistance were chosen as features. Then, the four classifiers were analysed using the Boruta package [20]. Four different methods were used to ensure validity of the results. Table 5 summarises the results of these methods for 4-year-old samples and Table 6 summarises the result for samples less than 18 months old. From Tables 5 and 6, it is clear that Sparse Logistic Regression worked more efficiently than the other methods since it has a greater area under curve value. Figures 13 and 14 indicate the ROC analysis of 128 samples of VOCs differentiating between diabetes and control samples for two different sample age groups.

Table 5. ROC with Boruta package analysis of data from 4-year-old samples.

Methods	AUC	Sensitivity	Specificity	PPV	NPV	*p*-Value
Sparse Logistic Regression	0.89 (0.83–0.95)	0.65 (0.53–0.76)	0.98 (0.89–1)	0.98	0.64	1.583×10^{-13}
Random Forest	0.89 (0.84–0.95)	0.69 (0.58–0.79)	0.9 (0.77–0.97)	0.91	0.65	1.088×10^{-13}
Gaussian Process	0.85 (0.78–0.92)	0.77 (0.66–0.86)	0.85 (0.72–0.94)	0.89	0.71	4.04×10^{-11}
Support Vector Machine	0.78 (0.69–0.88)	0.88 (0.78–0.94)	0.69 (0.54–0.81)	0.81	0.79	8.529×10^{-8}

Table 6. ROC with Boruta package analysis for data from 18-month-old samples.

Methods	AUC	Sensitivity	Specificity	PPV	NPV	*p*-Value
Sparse Logistic Regression	0.99 (0.96–1)	0.98 (0.89–1)	0.97 (0.86–1)	0.98	0.97	3.639×10^{-15}
Random Forest	0.97 (0.94 –1)	0.98 (0.89–1)	0.87 (0.72–0.96)	0.91	0.97	4.317×10^{-14}
Gaussian Process	0.94 (0.89–0.99)	0.9 (0.78–0.97)	0.89 (0.75–0.97)	0.92	0.87	9.162×10^{-13}
Support Vector Machine	0.94 (0.87–1)	0.98 (0.89–1)	0.89 (0.75–0.97)	0.92	0.97	9.733×10^{-13}

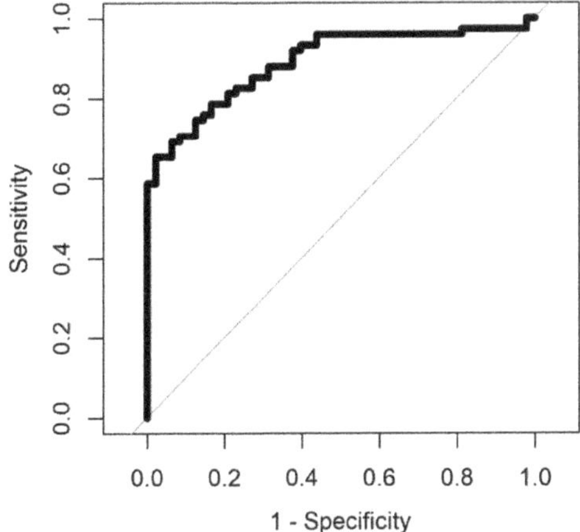

Figure 13. ROC with Boruta package analysis for data from samples up to 4 years old. Sparse Logistic Regression (AUC = 0.89).

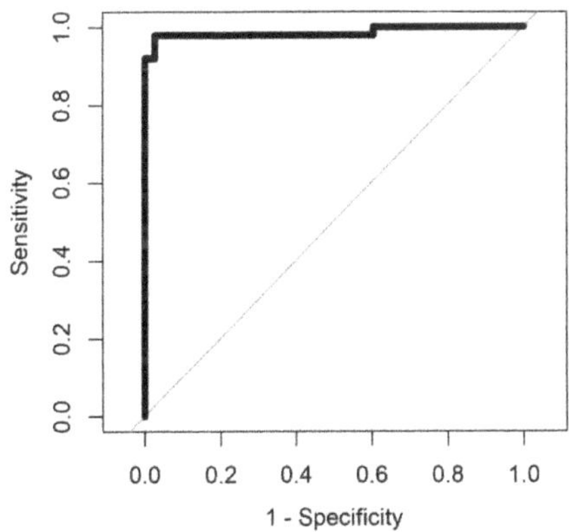

Figure 14. ROC with Boruta package analysis for data from samples less than 18 months old. Sparse Logistic Regression (AUC = 0.99).

4. Discussion

This paper has introduced diabetes as a major global health concern, affecting 1 in 12 of the population. The power of FAIMS and the FOX 4000 eNose to distinguish healthy and diabetic patients is considerable. Using an electronic nose, along with statistical and machine learning techniques, it was shown that we can accurately classify diabetic patients from healthy controls using only the aromas emanating from a urine sample. High prediction accuracy was achieved by combining PCA with a sparse logistic regression and a Gaussian process classifier. No single sensor was found to be able to distinguish healthy and disease patients, yet combining all sensors allows a high degree of predictive accuracy. It offers hope in developing a low-cost, point of care, rapid diagnostic tool that could potentially be an alternative non-invasive means to diagnose and, in the future, monitor the progression of diabetes.

The secondary result of this paper is proof that the vapours emanating from a stored urine sample are affected by storage time, as demonstrated by using the FOX 4000 and FAIMS electronic nose instruments. In most cases, samples would be tested well before this four-and-a-half-year period. Our study suggests that this is not feasible when dealing with urine samples for gas analysis. The results presented in this paper suggest that the optimal timing for urine analysis is less than 12 months and certainly not beyond this sample age.

5. Conclusions

Diabetes affects a large proportion of the world population and results in millions of deaths every year. Currently more than 40% of individuals with type 2 diabetes are undiagnosed. Here, an Alpha M.O.S FOX-4000 and FAIMS electronic nose were used to analyse urinary aromas from subjects with type 2 diabetes and healthy controls. By performing PCA and applying a classification algorithm, a high predictive accuracy was achieved. This study provides evidence suggesting that it may be possible to use urinary gas phase bio-markers to diagnose and monitor diabetes. Discriminating diabetic from control samples with above 95% accuracy proves that it is possible to diagnose diabetes from VOCs emitted from urine sample with eNose instruments. FAIMS can distinguish between diabetes and control samples that are less than a year old with sensitivity greater than 90% and specificity greater than 80%. FOX 4000 can separate diabetic from control samples with sensitivity and specificity above 90%. Also, it is suggested that samples under 12 months of age produce enough VOCs for urine analysis using an eNose.

Author Contributions: Conceptualization, J.A.C. and R.P.A.; Methodology, S.E., E.M., R.P.A. and J.A.C.; Data Collection, S.E. and E.M.; Data Analysis, A.W. and S.E.; Writing-Original Draft Preparation, S.E. and J.A.C.; Writing-Review & Editing, S.E., J.A.C., E.M., A.W. and R.P.A.

Funding: This work was supported by Indonesia Endowment Fund for Education (LPDP), Ministry of Finance, Republic of Indonesia.

Conflicts of Interest: The authors declare no conflict of interest.

References

1. Persaud, K.; Dodd, G. Analysis of discrimination mechanisms in the mammalian olfactory system using a model nose. *Nature* **1982**, *299*, 352–355. [CrossRef] [PubMed]
2. Westenbrink, E.; Arasaradnam, R.P.; O'Connell, N.; Bailey, C.; Nwokolo, C.; Bardhan, K.D.; Covington, J.A. Development and application of a new electronic nose instrument for the detection of colorectal Cancer. *Biosens. Bioelectron.* **2014**, *67*, 733–738. [CrossRef] [PubMed]
3. Covington, J.A.; Wedlake, L.; Andreyev, J.; Ouaret, N.; Thomas, M.G.; Nwokolo, C.U.; Bardhan, K.D.; Arasaradnam, R.P. The detection of patients at risk of gastrointestinal toxicity during pelvic radiotherapy by electronic nose and FAIMS: A pilot study. *Sensors* **2012**, *12*, 13002–13018. [CrossRef] [PubMed]
4. Van de Goor, R.; van Hooren, M.; Dingemans, A.-M.; Kremer, B.; Kross, K. Training and validating a portable electronic nose for lung cancer screening. *J. Thoracic Oncol.* **2018**, *13*, 676–681. [CrossRef] [PubMed]
5. Herman-Saffar, O.; Boger, Z.; Libson, S.; Lieberman, D.; Gonen, R.; Zeiri, Y. Early non-invasive detection of breast cancer using exhaled breath and urine analysis. *Comput. Biol. Med.* **2018**, *96*, 227–232. [CrossRef] [PubMed]
6. Kateb, B.; Ryan, M.A.; Homer, M.L.; Lara, L.M.; Yin, Y.; Higa, K.; Chen, M.Y. Sniffing out cancer using the JPL electronic nose: A pilot study of a novel approach to detection and differentiation of brain cancer. *NeuroImage* **2009**, *47*, T5–T9. [CrossRef] [PubMed]

7. Pennazza, G.; Santonico, M.; Bartolazzi, A.; Martinelli; Paolesse, R.; Di Natale, C.; Bono, R.; Tamburrelli, V.; Cristin, S.; D'Amico, A. Melanoma volatile fingerprint with a gas sensor array: In vivo and in vitro study. *Procedia Chem.* **2009**, *1*, 995–998. [CrossRef]

8. Bax, C.; Taverna, G.; Eusebio, L.; Sironi, S.; Grizzi, F.; Guazzoni, G.; Capelli, L. Innovative diagnostic methods for early prostate cancer detection through urine analysis: A review. *Cancers* **2018**, *10*. [CrossRef] [PubMed]

9. Sujono, H.A.; Rivai, M.; Amin, M. Asthma identification using gas sensors and support vector machine. *Telecommun. Comput. Electron. Control* **2018**, *16*, 1468–1480. [CrossRef]

10. Siyang, S.; Wongchoosuk, C.; Kerdcharoen, T. Diabetes diagnosis by direct measurement from urine odor using electronic nose. In Proceedings of the 5th 2012 Biomedical Engineering International Conference (BMEiCON), Ubon Ratchathani, Thailand, 5–7 December 2012. [CrossRef]

11. Mohamed, E.I.; Linder, R.; Perri, G.; Di Daniele, N.; Pöppl, S.J.; De Aorenzo, L. Predicting type 2 diabetes using an electronic nose-based artificial neural network analysis. *Diabetes Nutr. Metab.* **2002**, *15*, 215–221. [PubMed]

12. The International Diabetes Federation, Facts and Figures 2017. Available online: https://www.idf.org/aboutdiabetes/what-is-diabetes/facts-figures.html (accessed on 10 February 2018).

13. Diabetes, UK. Diabetes Prevalence 2017 (November 2017). Available online: https://www.diabetes.org.uk/professionals/position-statements-reports/statistics/diabetes-prevalence-2017 (accessed on 24 January 2018).

14. N.D.A. (NDA). National Diabetes Audit Complications and Mortality 2015–2016. 2017. Available online: https://digital.nhs.uk/catalogue/PUB30030 (accessed on 24 January 2018).

15. Esfahani, S.; Sagar, N.M.; Kyrou, I.; Mozdiak, E.; O'Connell, N.; Nwokolo, C.; Bardhan, K.D.; Arasaradnam, R.P.; Covington, J.A. Variation in Gas and Volatile Compound Emissions from Human Urine as It Ages, Measured by an Electronic Nose. *Biosensors* **2016**, *6*. [CrossRef] [PubMed]

16. Covington, J.A.; van der Schee, M.P.; Edge, A.S.L.; Boyle, B.; Savage, R.S.; Arasaradnam, R.P. The application of FAIMS gas analysis in medical diagnostics. *Analyst* **2015**, *140*, 6775–6781. [CrossRef] [PubMed]

17. Covington, J.A.; Westenbrink, E.W.; Ouaret, N.; Harbord, R.; Bailey, C.; O'Connell, N.; Cullis, J.; Williams, N.; Nwokolo, C.U.; Bardhan, K.D.; et al. Application of a novel tool for diagnosing bile acid diarrhea. *Sensors* **2013**, *13*, 11899–11912. [CrossRef] [PubMed]

18. Arasaradnam, R.P.; Ouaret, N.; Thomas, M.G.; Quraishi, N.; Heatherington, E.; Nwokolo, C.U.; Bardhan, K.D.; Covington, J.A. A novel tool for noninvasive diagnosis and tracking of patients with inflammatory bowel disease. *Inflamm. Bowel Dis.* **2013**, *19*, 999–1003. [CrossRef] [PubMed]

19. Lin, H.; Yan, Y.; Zhao, T.; Peng, L.; Zou, H.; Li, J.; Yang, X.; Xiong, Y.; Wang, M.; Wu, H. Rapid discrimination of Apiaceae plants by electronic nose coupled with multivariate statistical analyses. *J. Pharm. Biomed. Anal.* **2013**, *84*, 1–4. [CrossRef] [PubMed]

20. Kursa, M.B.; Rudnicki, W.R. Feature selection with the Boruta package. *J. Stat. Softw.* **2010**, *36*, 1–13. [CrossRef]

Article

Cerium Oxide-Tungsten Oxide Core-Shell Nanowire-Based Microsensors Sensitive to Acetone

Milena Tomić [1], Milena Šetka [2], Ondřej Chmela [2], Isabel Gràcia [1], Eduard Figueras [1], Carles Cané [1] and Stella Vallejos [1,2,*]

[1]   Instituto de Microelectrónica de Barcelona (IMB-CNM, CSIC), Campus UAB, 08193 Bellaterra, Spain; milena.tomic@imb-cnm.csic.es (M.T.); isabel.gracia@imb-cnm.csic.es (I.G.); Eduard.Figueras@imb-cnm.csic.es (E.F.); carles.cane@imb-cnm.csic.es (C.C.)
[2]   CEITEC-Central European Institute of Technology, Brno University of Technology, 61200 Brno, Czech Republic; milena.setka@ceitec.vutbr.cz (M.Š.); Ondrej.Chmela@ceitec.vutbr.cz (O.C.)
*   Correspondence: stella.vallejos@imb-cnm.csic.es; Tel.: +34-935-947700

Received: 31 October 2018; Accepted: 19 November 2018; Published: 23 November 2018

Abstract: Gas sensitive cerium oxide-tungsten oxide core-shell nanowires are synthesized and integrated directly into micromachined platforms via aerosol assisted chemical vapor deposition. Tests to various volatile organic compounds (acetone, ethanol, and toluene) involved in early disease diagnosis demonstrate enhanced sensitivity to acetone for the core-shell structures in contrast to the non-modified materials (i.e., only tungsten oxide or cerium oxide). This is attributed to the high density of oxygen vacancy defects at the shell, as well as the formation of heterojunctions at the core-shell interface, which provide the modified nanowires with 'extra' chemical and electronic sensitization as compared to the non-modified materials.

Keywords: gas sensors; volatile organic compounds (VOCs); acetone; metal oxides; heterojunctions

---

1. Introduction

Odor (gas, vapor, aroma) detection systems are of high interest as they are non-invasive key-enabling technologies, which are relevant in both traditional (e.g., environment, industry) and innovative applications such as the early detection of diseases from exhaled breath [1,2]. The literature related to exhaled breath as bio information for disease diagnosis has shown previously that human breath contains more than two hundred different gases and volatile organic compounds (VOCs) species that vary from person to person [3]. In the midst of a wide range of analytes, acetone, toluene, and ethanol are within the most relevant VOCs that are typically involved in various diseases, including diabetes and cancer. Thus, for instance, high acetone concentrations (~21 ppm) have been identified in the exhaled breath of diabetic patients, as compared to healthy patients (~2.7 ppm) and oppositely to patients with lung cancer, who showed lower acetone concentrations (~0.9 ppm). Similarly, the ethanol concentration in breath has shown an increase (~2.1 ppm) in patients with lung/breast cancer and diabetes with respect to healthy patients (~0.2 ppm) [4–6].

Currently, the analysis of breath is still an emerging diagnosis technology that uses large and expensive laboratory equipment such as gas-chromatograph, ion-mobility and/or mass spectrometers [7]. In the future, however, miniaturized, portable, and wearable systems with enhanced functionality (sensitivity, selectivity, and stability) and high autonomy at a low cost could substitute this equipment; a fact that demands the innovation of current odor detection systems. In this context, metal oxide (MOX) gas sensors based on the chemoresistive principle represent an alternative to bulky equipment providing simpler architecture and fabrication processes compatible with 'standard' MEMS and CMOS technologies [2]. Chemoresistive gas sensors devices generally consist of a transducing platform (microscale) and a gas sensitive MOX optimized to interact with specific groups of gaseous or vapor analytes. However, among the diverse issues that may potentially be addressed to improve the functionality of these monitoring systems (e.g., optimization of sensing modes and control electronics or the integration of smart systems with new micro/nano fabrication concepts), the focus on nanoscaled sensitive materials is still essential to radically improve their performance. Thus, several studies have demonstrated that MOXs modified with second-phase constituents, either nanosized noble metals or other MOXs, have a positive effect on the sensing properties of both the host MOX and the second-phase constituent, particularly when the size of both materials is within the Debye length of the surface (typically on the order of $2-100$ nm) [8]. Moreover, recently, it has been pointed out that the modification of a MOX with noble metals or other MOXs allows for the formation of nanoscale heterojunctions and, in turn, sensing mechanisms dominated not only by the surface, but also the interface, which has proved to improve the sensing properties of these materials [9].

In recent years, tungsten oxide has demonstrated high potential in gas sensing among traditional gas sensitive MOXs such as SnO_2 and ZnO (Figure 1), showing a strong sensitivity to oxidizing gases including nitrogen dioxide and ozone [10]. Moreover, the modification of tungsten oxide with second-phase constituents such us platinum, copper oxide, or iron oxide has shown an improved sensitivity and selectivity to reducing species including hydrogen [11], hydrogen sulfide [12] or toluene [13], respectively. As far as cerium oxide is concerned, the peculiarity that makes this oxide also attractive in gas sensors reside overall in the defect sites caused by the valence state changes between Ce^{4+} and Ce^{3+}, which considerably alter the concentration of oxygen vacancies, and provides a good redox behavior and catalytic activity [14–16]. However, and despite these favorable surface properties the use of cerium oxide in gas sensing is still infrequent (Figure 1) and its sensing properties upon VOCs have not been fully explored in the literature related to gas sensors.

Optimized gas sensitive MOXs need synthetic methods able to produce well defined and even structures. Additive (bottom-up) synthetic methods, as opposed to subtractive (top-down) methods, are ideal for this task and industrially attractive as they provide the ability to generate films in a continuous mode with high purities and high throughput. Aerosol assisted (AA) chemical vapor deposition (CVD) is a versatile additive synthetic method used previously to obtain non-modified (e.g., WO_3) or metal/MOX modified MOXs (e.g., Pt/WO_3, Fe_2O_3/WO_3) [11,13]. Additionally, recently, the AACVD of cerium oxide from Ce (dbm)$_4$ has been proved as a strategy to overcome the low volatility of traditional cerium CVD precursors [17]. In this work, however, we achieve the AACVD of cerium oxide from Ce (acac)$_3$ precursor and use this route to synthesize cerium oxide-tungsten oxide core-shell nanowires in a two-step process performed directly on silicon-based micromachined platforms. In addition, we validate the sensing properties of these systems to acetone, and other relevant VOCs monitored in early disease diagnosis.

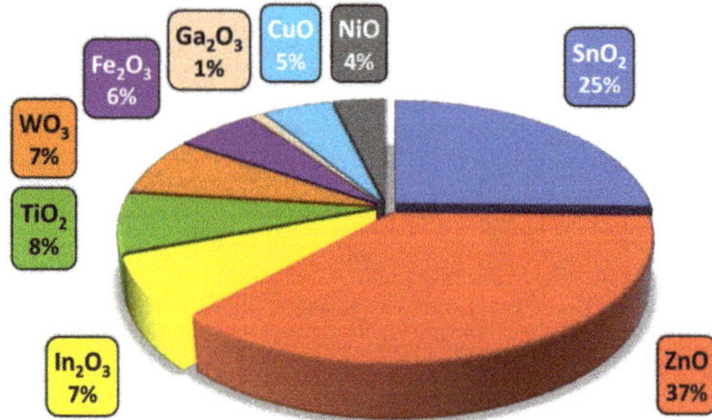

Figure 1. A survey of the most applicable gas sensitive MOXs reported in the literature (Web of Science database from 1998 to 2018).

2. Materials and Methods

Tungsten oxide (non-modified nanowires), cerium oxide (non-modified porous films), and cerium oxide-tungsten oxide core-shell nanowires were grown directly onto micromachined transducing platforms (Figure 2a) [18] using the AACVD system described previously [19]. AACVD is a variant of the conventional CVD technique, which uses aerosol to transport dissolved precursors to a heated reaction zone. Here, the non-modified tungsten oxide nanowires films were deposited at 350 °C from a solution of tungsten hexacarbonyl (30 mg, $W(CO)_6$, Sigma-Aldrich, St. Louis, MO, USA, $\geq$97%) and methanol (5 mL Sigma-Aldrich, $\geq$99.6%), whereas the non-modified cerium oxide films were deposited at 500 °C from cerium (III) acetylacetonate hydrate (28 mg, $Ce\,(acac)_3 \cdot H_2O$, Sigma-Aldrich) dissolved in methanol (2 ml, Sigma-Aldrich). On the other hand, the cerium oxide-tungsten oxide core-shell nanowires were achieved using a two-step AACVD process [20], in which the tungsten oxide nanowire cores were deposited at 350 °C in the first step and the cerium oxide shell film at 500 °C in the second step employing the same protocols described above for the non-modified films. Finally, the non-modified and modified films were annealed at 500 °C in air.

The morphology of the films was examined using scanning electron microscopy (SEM— Auriga Series, 3 KV, Carl Zeiss, Jena, Germany) and the phase using X-ray Diffraction (XRD–Bruker-AXS, model A25 D8 Discover, Cu $K\alpha$ radiation, Billerica, MA, USA). Further analysis of the material was carried out using X-ray photoelectron spectroscopy (XPS—Kratos Axis Supra with monochromatic Al $K\alpha$ X-ray radiation, an emission current of 15 mA and hybrid lens mode, Manchester, UK). The survey and detailed spectra were measured using pass energy of 80 eV and 20 eV, respectively. The band gap of the films was estimated by measuring the diffuse reflectance (AvaSpec-UV/VIS/NIR, Avantes, Apeldoorn, the Netherlands) of the films and performing Kubelka–Munk transformation.

The microsensors were tested in a continuous flow test chamber provided with mass flow controllers that allow the mixture of dry/humid air and calibrated gaseous analytes (ethanol, acetone, toluene, carbon monoxide and hydrogen purchased from Praxair, Danbury, CT, USA) to obtain the desired concentration. To have a proper control of the relative humidity (RH) inside the gas test chamber, an evaluation kit (EK-H4, Sensirion AG, Stäfa, Switzerland) with a humidity sensor was also used. The dc resistance measurements of the microsensor were achieved in a system provided with an electrometer (Keathley 6517A, Cleveland, OH, USA) and a multimeter (Keathley 2700, Cleveland, OH, USA) with switch system to monitor various sensors simultaneously. More details of the characterization systems were reported elsewhere [18]. The sensor response was defined as Ra/Rg, where Ra and Rg are the resistance in dry/humid air and the resistance after 600 s of analyte exposure, respectively. The sensors were tested for a period of one month during which each sensor accumulated 180 h of operation under the different conditions (analytes, temperatures, humidity) employed.

3. Results

3.1. Gas Sensitive Films

SEM imaging of the microsensors after AACVD of the gas sensitive structures showed uniform deposited films that covered the electrodes integrated into the micromachined membrane (Figure 2a). A close view of the non-modified tungsten oxide wires (*W*) showed bare and even surfaces as noticed previously for other AACVD tungsten oxide structures [21]. In contrast, a close view of the cerium oxide-tungsten oxide core-shell wires (*Ce/W*) displayed the presence of a rugged thin film covering the wire surface (Figure 2b,c), similarly to that observed when depositing non-modified cerium-based films (*Ce*) from a $Ce\,(acac)_3$ methanolic solution via AACVD (Figure 2d).

Generally, AACVD of the non-modified (*W* and *Ce*) and modified (*Ce/W*) films showed a good adherence to the substrate, with the wire-like morphology films (i.e., *W* and *Ce/W*) forming a mat-like network of non-aligned nanowires with diameters below 100 nm, and the particle-like morphology films (i.e., *Ce*) displaying a porous surface composed of grains with diameters below 40 nm. The as-deposited non-modified *W* films displayed a bluish color, whereas the *Ce/W* films displayed a dark yellowish to dark green color, similarly to the color observed on the *Ce* films. However, after annealing the non-modified *W* and modified *Ce/W* films became whitish and pale yellowish, respectively. Figure 3 displays the diffuse reflectance spectra of the films deposited without modification (i.e., only tungsten oxide or only cerium oxide) via AACVD. These measurements and their corresponding Kubelka–Munk transformation indicated optical band gaps at ~3.2 eV for the tungsten oxide films and ~3.1 eV for the cerium oxide films, in agreement with the literature band gap values of tungsten oxide (2.6–3.7 eV) [22] and cerium oxide (2.7–3.4 eV) [15].

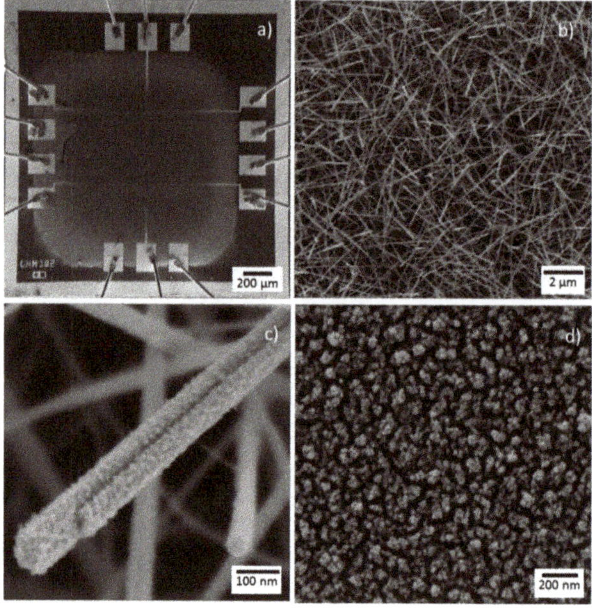

Figure 2. The SEM imaging of a (**a**) gas sensor device, the *Ce/W* nanowires at (**b**) low and (**c**) high magnification, and (**d**) the non-modified *Ce* films integrated on the micromachined membrane.

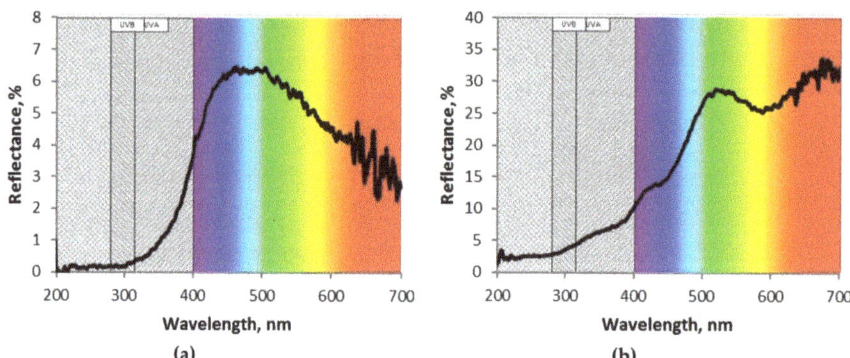

Figure 3. The diffuse reflectance spectra of the aerosol assisted chemical vapor deposited (**a**) tungsten oxide and (**b**) cerium oxide films without modification.

XRD analysis of the films revealed the presence of a monoclinic-phase (International Centre of Diffraction Data–ICDD card no. 72-0677) in the *W* and *Ce/W* films with greatly enhanced intensity (preferred orientation) in the [001] direction, consistent with our previous results for AACVD of tungsten oxide [13]. A weak diffraction peak was also noticed at 47.8° 2θ for the *Ce/W* films

(Figure 4). This diffraction peak is in line with the pattern identified on the non-modified *Ce* based films corresponding to cerium dioxide (Crystallography Open Database–COD ID card no. 7217887).

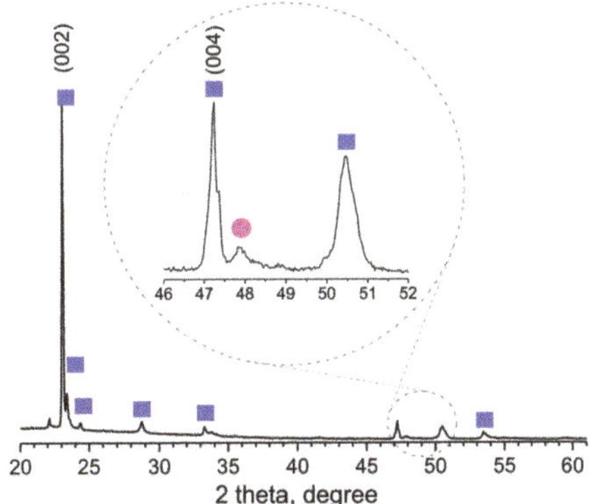

Figure 4. The XRD pattern of the *Ce/W* films. The diffraction peak at 47.8° 2θ (full pink circle) corresponds to cerium dioxide cubic phase (P1), COD ID card no. 7217887, the rest of the diffraction peaks (full blue squares) in the data can be indexed to a monoclinic phase (P21/n), ICDD card no. 72-0677, with only peaks of greatly enhanced intensity (preferred orientation), specifically indexed.

The XPS of both *W* and *Ce/W* films exhibited typical W $4f_{7/2}$, W $4f_{5/2}$ and W $5p_{3/2}$ XPS core level peaks (Figure 5a), consistent with the literature and previous tungsten oxide nanowires synthetized via AACVD [13]. XPS narrow scan spectra of the Ce 3d core level peaks at the *Ce* and *Ce/W* wires displayed multiplet splitting between 875 and 920 eV in agreement with the standard binding energies for Ce 3d peaks and partially reduced cerium oxide [23,24]. Figure 5b displays the experimental data and the corresponding deconvolution of the Ce 3d spectrum recorded on the *Ce/W* wires. The peaks v, v'' and v''' are attributed to the main and satellite peaks of the Ce^{4+} state, whereas the peaks v_0, v' correspond to the peaks of Ce^{3+} state. The relative contribution of Ce^{4+} and Ce^{3+} species at the *Ce/W* films was estimated from the ratio of integrated Ce^{4+} peaks to the total Ce^{4+} and Ce^{3+} peaks, finding a value of ~42% for Ce^{4+} and 58% Ce^{3+} species. The relatively high amount of Ce^{3+} species indicate a charge imbalance with oxygen vacancy defects and an unsaturated chemical bond at the *Ce/W* film suggesting a high redox nature of the film.

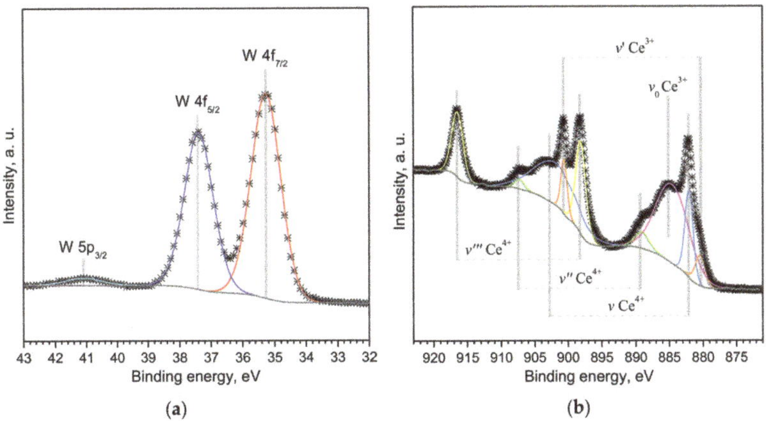

Figure 5. (**a**) The W 4f and (**b**) Ce 3d spectra recorded on the cerium-modified wires. The W 4f spectrum recorded on the non-modified tungsten oxide wires showed similar characteristics.

Linear extrapolation of the valence band (VB) leading edge on the XPS spectra recorded on the *Ce/W* film near the Fermi level ($E_B = 0$) indicates the simultaneous presence of both cerium oxide and tungsten oxide induced VB (Figure 6a). One can notice that the VB onset for cerium oxide occurs ~0.5 eV (ΔE_V) above the VB onset for tungsten oxide. Therefore, according to the band gap estimated by diffuse reflectance for each non-modified material in the position of the conduction band (CB) of cerium oxide is ~0.4 eV (ΔE_C) above the CB of tungsten oxide, consequently suggesting a staggered type of heterojunction at the interface of the *Ce/W* core-shell structures (Figure 6b). In contrast, the linear extrapolation of VB leading edge on the XPS spectra of the *W* and *Ce* films showed only the presence of tungsten oxide induced VB at 2.9 eV in agreement with previous reports [13].

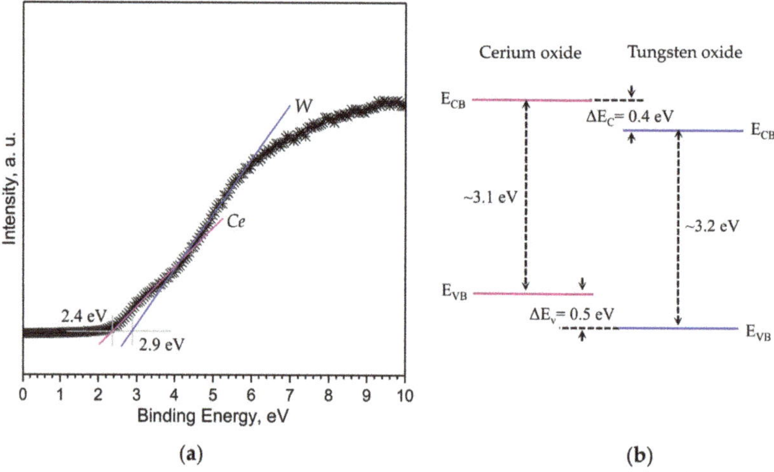

Figure 6. (**a**) The XPS valence band spectra of the cerium-modified tungsten oxide nanowires, and (**b**) schematic of the estimated energy level diagram at the interface. E_{CB} and E_{VB} represent the conduction band minimum and the valence band maximum (not to scale).

In brief, the characterization of the films deposited via AACVD demonstrates the formation of crystalline cerium oxide-tungsten oxide core-shell wires with a relatively high amount of Ce^{3+} species at the surface and the presence of the characteristic valence band onsets for cerium oxide and tungsten oxide.

3.2. Gas Sensing Tests

Overall, the sensors displayed an n-type response with a good reproducibility to the target gaseous analytes (acetone, ethanol, toluene, carbon monoxide, and hydrogen) and relatively low variations of the baseline resistance along the testing period. Gas sensing tests of the microsensors were carried out at various operating temperatures from 150 °C to 400 °C by DC resistance measurements of the films integrated via AACVD. These test proved a better sensor functionality to 80 ppm of acetone at 300 °C for the sensors based on *W* and *Ce/W* films and 400 °C for the sensors based on *Ce* films. As the *Ce* based sensors registered comparatively higher baseline resistances (~40 GΩ at 400 °C) than *W* (13 kΩ at 300 °C) and *Ce/W* (65 kΩ at 300 °C) sensors, additionally requiring higher temperatures to achieve the maximum responses (e.g., response to acetone 4.9 at 400 °C for the *Ce* sensors and 5.6 at 300 °C for the *Ce/W* sensors), further analyses related to the analyte concentration and humidity dependence of the sensor response were performed only for the *W* and *Ce/W* based sensors at 300 °C.

Figure 7a displays the sensor response to 80 ppm of each analyte and type of sensor at 300 °C. These results show the improved responses registered for the *Ce/W* sensors, as opposed to the *W* sensors, as well as the higher responses to acetone compared to the rest of the analytes. Results in

Figure 7a also suggests an improved selectivity for the *Ce/W* films with lower cross-responses among the analytes; for instance, the difference of the response to acetone in relation to ethanol is 1.6 for the *Ce/W* films and 1 for the *W* films. The low cross response registered on the *Ce/W* sensor is noticed in more detail in Figure 7b, in which is displayed the analysis of variance (ANOVA) realized for a data set comprising four replicates for each type of sensor and analyte. Additionally, the principal component (PC) analysis performed using replicated responses of both sensors (i.e., *W* and *Ce/W*) to each analyte is represented in Figure 8. These results, specifically the scores, which correspond to the projections of the measurements in an orthogonal base of PCs, indicate the possibility to improve the discrimination of the analytes by using an array of *W* and *Ce/W* based sensors.

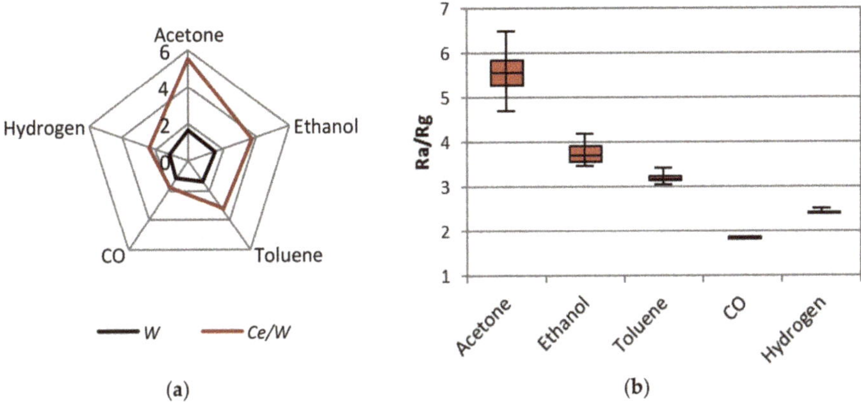

Figure 7. (a) The radial plot of the sensor response to 80 ppm of acetone, ethanol, toluene, carbon monoxide, and hydrogen using the *W* and *Ce/W* based sensors. (b) Box plots of the sensor response to each analyte recorded by the *Ce/W* based sensors. Each box displays the median and upper and lower quartiles (first and third) of the respective distribution. Box whiskers indicate the dispersion of the measurements.

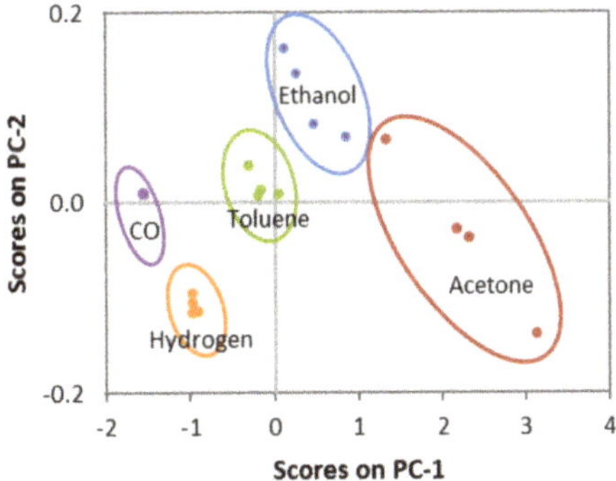

Figure 8. Principal component analysis applied to discriminate the tested VOCs by using an array of non-modified tungsten oxide wires and the cerium oxide-tungsten oxide core-shell wire-based sensors.

Further tests of the sensors to various concentrations of each analyte showed direct proportional changes in the response to concentration. An example of the response registered with both types of sensors to acetone is shown in Figure 9. For these conditions, the limit of detection corresponding to three times the noise level [25] was estimated at 1 ppm for *W* and 0.2 ppm for *Ce/W* sensors. Overall, the changes respect to concentration for the *W* sensors proved a lower sensitivity compared to the *W/Ce* sensors, which demonstrates a better sensitivity to the analytes. The sensitivity (S), defined as the ratio between the change in response (ΔR) and a fixed change in analyte concentration (ΔC) for each sensor and analyte, was registered to be nearly five times higher for acetone and three times higher for ethanol when using the cerium-modified sensors (found $\Delta R/\Delta C_{acetone}$ for *W* sensors 0.8%, *Ce/W* sensors 4.7%; $\Delta R/\Delta C_{ethanol}$ for *W* sensors 0.8%, *Ce/W* sensors 2.2%).

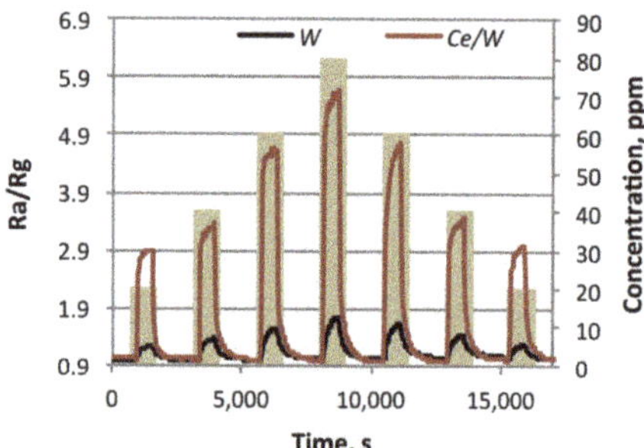

Figure 9. The sensor response to various concentrations (from 20 to 80 ppm) of acetone recorded with the *W* and *Ce/W* based sensors.

Additional tests of the sensors in a controlled humid ambient (10 and 20% RH), consistent with those reported after preconditioning the relative humidity in breath samples [26], registered lower sensor response to the analytes. The loss of response in humid ambient is a consequence of the proportional drop of the baseline resistance to relative humidity. This proportional change is usually present in metal oxides exposed to humidity due to the formation of hydroxyl groups at temperatures above 100 °C [18]. Currently, most of the strategies to attenuate further the humidity interference from the material point of view are connected with the fine tune of the MOX morphology [27] and/or the incorporation of humidity-insensitive additives (e.g., NiO [28], CuO [12], or SiO_2 [29]). Figure 10 displays the typical resistance changes for the *W* and *Ce/W* sensors to each mixture of RH and acetone tested and their replicates.

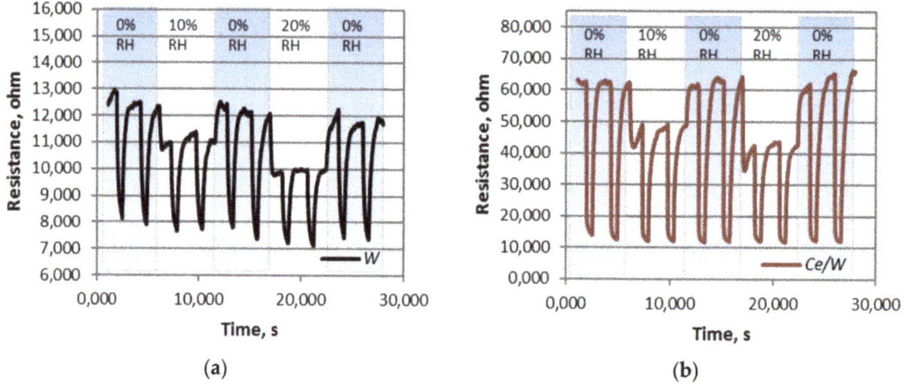

Figure 10. The resistance changes registered for (**a**) *W* and (**b**) *Ce/W* to acetone in dry and humid ambient (RH: relative humidity).

Previous reports in the literature related to acetone sensing using modified tungsten oxide films (with Au, Pd, AuPd) suggest the functionality of tungsten oxide at 300 °C for relatively high acetone concentrations (200–1000 ppm) [30]. Further tests, also performed on tungsten oxide, modified with TiO_2 [31] or Si [32] show the functionality of the sensors at similar (30 ppm) or lower (100–600 ppb) acetone concentrations, respectively, although requiring higher operating temperatures (400–500 °C), than those needed in this work. On the other hand, the use of cerium oxide as the gas sensitive element has been rarely reported in the literature, with the performance of this material having indicated the potential for sensing VOCs including acetone [33,34]. In general, the responses of the above mentioned non-miniaturized acetone sensors in the literature [30–34] are in the same order of magnitude than our micromachined sensors based on the cerium oxide-tungsten oxide core-shell wires, which suggests the viability of sensor miniaturization without losing the sensitivity of the system when optimizing the sensitive material. In addition, the good reproducibility of the responses during the testing period and the analysis (SEM, XRD) of the samples after the gas sensing experiments (which showed unchanged properties of the material with respect to the properties recorded initially) indicated a good stability of the sensors.

The enhanced functionality recorded on the *Ce/W* sensors is connected with the formation of heterojunctions at the interface of the tungsten oxide wires and the cerium oxide porous films. These heterojunctions are present due to the different band energies in both MOXs (Figure 6b), which facilitate electron migration from the cerium oxide film to the tungsten oxide wire. Generally, as oxygen is preadsorbed at the sensitive film during air exposure, the surface depletion region (L_D) and, in turn, the conduction channel along the wires are narrowed, which leads to lower conductivity along the film (Figure 11a). Alternatively, when a reducing analyte (i.e., VOCs) reacts with the preadsorbed oxygen, electrons are released back to the conduction band, and the depth of the surface depletion is narrowed, increasing the conduction channel along the wire and, in turn, its conductivity (Figure 11b). This mechanism controlled by the pre-adsorbed oxygen is similar for the non-modified and modified films, with the peculiarity that the charge transfer process (electron migration) occurring at the junction of the cerium oxide film and tungsten oxide wire provides larger electron density to the wires (accumulation layer) in the pre-adsorption cycle (as opposed to the non-modified *W* or *Ce* films). This allows for larger changes of the depletion layer and an enhanced modulation of the wire conduction channel, which is reflected finally on the sensor response.

In the same line, the lowering of the baseline resistance upon humidity implies a diminution of the chemically active oxygen species at the surface and, in turn, a narrow surface depletion in the air (pre-adsorption). Thus, as the conduction channel in humid ambient is wider than in dry ambient (Figure 11), the conduction changes induced by the reducing gas are less significant, and thus the sensor response is as well.

Summarizing, the gas sensing tests showed improved acetone sensing properties for *Ce/W* based microsensors showing a higher response, and better sensitivity and selectivity to the analytes tested in relation to the *W* or *Ce* based microsensors.

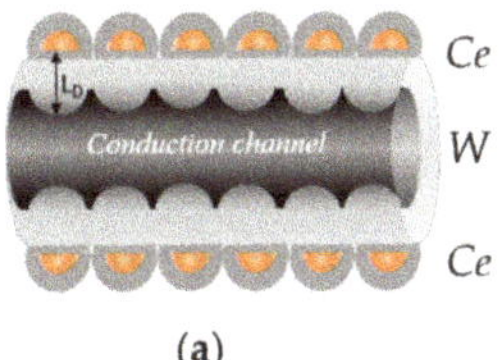

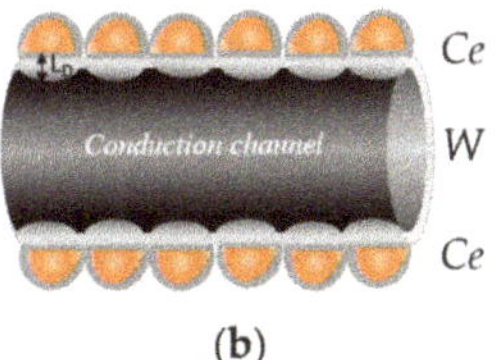

(a) (b)

Figure 11. The schematic view of the heterojunction formed at the surface of the cerium oxide (*Ce*) core-shell tungsten oxide (*W*) wires and the possible mechanism (**a**) after exposure to air and (**b**) reducing gases such as acetone. L_D is the Debye length or depth of the depletion region from the surface (not at scale).

4. Conclusions

These results demonstrate the formation of cerium oxide-tungsten oxide core-shell nanowires with improved response and sensitivity to acetone as compared to non-modified tungsten oxide wires or cerium oxide porous films. Tests of these sensors to acetone in humid ambient showed a drop of the responses as a consequence of the lowering of the baseline resistance due to humidity. Principal component analysis of the responses obtained for each analytes using an array the non-modified and modified sensitive films indicated the possibility to enhance the selectivity of the microsensors by improving the discrimination of analytes. The improved sensitivity is attributed to the formation of heterojunctions at the interface of both oxides (i.e., tungsten oxide and cerium oxide) which leads to an 'extra' chemical and electronic sensitization to the modified films as compared to the non-modified films.

Author Contributions: M.T.; synthesis and integration of gas sensitive films, M.S. and S.V.; XPS analysis, M.T. and I.G. diffuse reflectance tests, I.G., E.F. and C.C.; Si-based micromachined platforms; I.G., E.F. and O.C. gas sensing characterization, M.T., M.S. and O.C., literature search, M.T., I.G. and S.V.; writing review and editing, S.V. and C.C.; funding acquisition.

Funding: This work has been supported by the Czech Science Foundation (GAČR) via Grant no. 17-16531S, and the Spanish Ministry of Science and Innovation via the Ramón y Cajal Programme and Grants TEC2015-74329-JIN-(AEI/FEDER,EU) and TEC2016-79898-C6-(AEI/FEDER,EU). This research has used the Spanish ICTS Network MICRONANOFABS (partially funded by MINECO). CEITEC Nano Research Infrastructure (IDLM2015041, MEYS CR, 2016–2019) is also acknowledged.

Conflicts of Interest: The authors declare no conflict of interest.

References

1. Brattoli, M.; de Gennaro, G.; de Pinto, V.; Loiotile, A.D.; Lovascio, S.; Penza, M. Odour detection methods: Olfactometry and chemical sensors. *Sensors* **2011**, *11*, 5290–5322. [CrossRef] [PubMed]

2. Di Natale, C.; Paolesse, R.; Martinelli, E.; Capuano, R. Solid-state gas sensors for breath analysis: A review. *Anal. Chim. Acta* **2014**, *824*, 1–17. [CrossRef] [PubMed]

3. Phillips, M.; Herrera, J.; Krishnan, S.; Zain, M.; Greenberg, J.; Cataneo, R.N. Variation in volatile organic compounds in the breath of normal humans. *J. Chromatogr. B Biomed. Sci. Appl.* **1999**, *729*, 75–88. [CrossRef]

4. Das, S.; Pal, S.; Mitra, M. Significance of Exhaled Breath Test in Clinical Diagnosis: A Special Focus on the Detection of Diabetes Mellitus. *J. Med. Biol. Eng.* **2016**, *36*, 605–624. [CrossRef] [PubMed]

5. Dent, A.G.; Sutedja, T.G.; Zimmerman, P.V. Exhaled breath analysis for lung cancer. *J. Thorac. Dis.* **2013**, *5*, S540–S550. [CrossRef] [PubMed]

6. Li, J.; Peng, Y.; Liu, Y.; Li, W.; Jin, Y.; Tang, Z.; Duan, Y. Investigation of potential breath biomarkers for the early diagnosis of breast cancer using gas chromatography–mass spectrometry. *Clin. Chim. Acta* **2014**, *436*, 59–67. [CrossRef] [PubMed]

7. Van de Kant, K.D.; van der Sande, L.J.; Jöbsis, Q.; van Schayck, O.C.; Dompeling, E. Clinical use of exhaled volatile organic compounds in pulmonary diseases: A systematic review. *Respir. Res.* **2012**, *13*. [CrossRef] [PubMed]

8. Yamazoe, N. New approaches for improving semiconductor gas sensors. *Sens. Actuator B Chem.* **1991**, *5*, 7–19. [CrossRef]

9. Miller, D.R.; Akbar, S.A.; Morris, P.A. Nanoscale metal oxide-based heterojunctions for gas sensing: A review. *Sens. Actuator B Chem.* **2014**, *204*, 250–272. [CrossRef]

10. Vallejos, S.; Khatko, V.; Calderer, J.; Gracia, I.; Canè, C.; Llobet, E.; Correig, X. Micro-machined WO_3-based sensors selective to oxidizing gases. *Sens. Actuator B Chem.* **2008**, *132*, 209–215. [CrossRef]

11. Vallejos, S.; Umek, P.; Stoycheva, T.; Annanouch, F.; Llobet, E.; Correig, X.; De Marco, P.; Bittencourt, C.; Blackman, C. Single-step deposition of Au- and Pt-nanoparticle-functionalized tungsten oxide nanoneedles synthesized via aerosol-assisted CVD, and used for fabrication of selective gas microsensor arrays. *Adv. Funct. Mater.* **2013**, *23*, 1313–1322. [CrossRef]

12. Annanouch, F.E.; Haddi, Z.; Vallejos, S.; Umek, P.; Guttmann, P.; Bittencourt, C.; Llobet, E. Aerosol-assisted CVD-grown WO_3 nanoneedles decorated with copper oxide nanoparticles for the selective and humidity-resilient detection of H_2S. *ACS Appl. Mater. Interfaces* **2015**, *7*, 6842–6851. [CrossRef] [PubMed]

13. Vallejos, S.; Gràcia, I.; Figueras, E.; Cané, C. Nanoscale heterostructures based on $Fe_2O_3@WO_{3-x}$ nanoneedles and their direct integration into flexible transducing platforms for toluene sensing. *ACS Appl. Mater. Interfaces* **2015**, *7*, 18638–18649. [CrossRef] [PubMed]

14. Ma, R.; Jahurul Islam, M.; Amaranatha Reddy, D.; Kim, T.K. Transformation of CeO_2 into a mixed phase CeO_2/Ce_2O_3 nanohybrid by liquid phase pulsed laser ablation for enhanced photocatalytic activity through Z-scheme pattern. *Ceram. Int.* **2016**, *42*, 18495–18502. [CrossRef]

15. Montini, T.; Melchionna, M.; Monai, M.; Fornasiero, P. Fundamentals and Catalytic Applications of CeO_2-Based Materials. *Chem. Rev.* **2016**, *116*, 5987–6041. [CrossRef] [PubMed]

16. Magesh, G.; Viswanathan, B.; Viswanath, R.P.; Varadarajan, T.K. Photocatalytic behavior of CeO_2-TiO_2 system for the degradation of methylene blue. *Indian J. Chem. Sect. A* **2009**, *48A*, 480–488.

17. Evans, M.; Di Maggio, F.; Blackman, C.; Sankar, G. AACVD synthesis of catalytic gold nanoparticle-modified cerium(IV) oxide thin films. *Phys. Status Solidi C* **2015**, *12*, 996–1000. [CrossRef]

18. Vallejos, S.; Grácia, I.; Chmela, O.; Figueras, E.; Hubálek, J.; Cané, C. Chemoresistive micromachined gas sensors based on functionalized metal oxide nanowires: Performance and reliability. *Sens. Actuator B Chem.* **2016**, *235*, 525–534. [CrossRef]

19. Vallejos, S.; Pizúrová, N.; Čechal, J.; Gràcia, I.; Cané, C. Aerosol-assisted chemical vapor deposition of metal oxide structures: Zinc oxide rods. *J. Vis. Exp.* **2017**, *127*, 56127. [CrossRef] [PubMed]

20. Annanouch, F.E.; Haddi, Z.; Ling, M.; Di Maggio, F.; Vallejos, S.; Vilic, T.; Zhu, Y.; Shujah, T.; Umek, P.; Bittencourt, C.; et al. Aerosol-Assisted CVD-Grown PdO Nanoparticle-Decorated Tungsten Oxide Nanoneedles Extremely Sensitive and Selective to Hydrogen. *ACS Appl. Mater. Interfaces* **2016**, *8*, 10413–10421. [CrossRef] [PubMed]

21. Vallejos, S.; Gràcia, I.; Figueras, E.; Cané, C. Catalyst-free vapor-phase method for direct integration of gas sensing nanostructures with polymeric transducing platforms. *J. Nanomaterials* **2014**, *2014*. [CrossRef]

22. Watanabe, H.; Fujikata, K.; Oaki, Y.; Imai, H. Band-gap expansion of tungsten oxide quantum dots synthesized in sub-nano porous silica. *Chem. Comm.* **2013**, *49*, 8477–8479. [CrossRef] [PubMed]

23. Mysliveček, J.; Matolín, V.; Matolínová, I. Heteroepitaxy of Cerium Oxide Thin Films on Cu(111). *Materials* **2015**, *8*, 6346–6359. [CrossRef] [PubMed]

24. Naganuma, T.; Traversa, E. Stability of the Ce^{3+} valence state in cerium oxide nanoparticle layers. *Nanoscale* **2012**, *4*, 4950–4953. [CrossRef] [PubMed]

25. Nicolas, J.; Romain, A.-C. Establishing the limit of detection and the resolution limits of odorous sources in the environment for an array of metal oxide gas sensors. *Sens. Actuator B Chem.* **2004**, *99*, 384–392. [CrossRef]

26. Prabhakar, A.; Iglesias, R.A.; Shan, X.; Xian, X.; Zhang, L.; Tsow, F.; Forzani, E.S.; Tao, N. Online Sample Conditioning for Portable Breath Analyzers. *Anal. Chem.* **2012**, *84*, 7172–7178. [CrossRef] [PubMed]

27. Vallejos, S.; Gràcia, I.; Pizúrová, N.; Figueras, E.; Hubálek, J.; Cané, C. Tuning of the Humidity-Interference in Gas Sensitive Columnar ZnO Structures. *Proceedings* **2017**, *1*. [CrossRef]

28. Wang, J.; Yang, P.; Wei, X. High-Performance, Room-Temperature, and No-Humidity-Impact Ammonia Sensor Based on Heterogeneous Nickel Oxide and Zinc Oxide Nanocrystals. *ACS Appl. Mater. Interfaces* **2015**, *7*, 3816–3824. [CrossRef] [PubMed]

29. Niarchos, G.; Dubourg, G.; Afroudakis, G.; Georgopoulos, M.; Tsouti, V.; Makarona, E.; Crnojevic-Bengin, V.; Tsamis, C. Humidity Sensing Properties of Paper Substrates and Their Passivation with ZnO Nanoparticles for Sensor Applications. *Sensors* **2017**, *17*. [CrossRef] [PubMed]

30. Kim, S.; Park, S.; Park, S.; Lee, C. Acetone sensing of Au and Pd-decorated WO_3 nanorod sensors. *Sens. Actuator B Chem.* **2015**, *209*, 180–185. [CrossRef]

31. Bertuna, A.; Comini, E.; Poli, N.; Zappa, D.; Sberveglieri, G. Acetone Detection by Chemical Sensors Based on Tungsten and Titanium Oxide Nanowires. *Proceedings* **2017**, *1*. [CrossRef]

32. Righettoni, M.; Tricoli, A.; Pratsinis, S.E. Si:WO_3 Sensors for Highly Selective Detection of Acetone for Easy Diagnosis of Diabetes by Breath Analysis. *Anal. Chem.* **2010**, *82*, 3581–3587. [CrossRef] [PubMed]

33. Pandeeswari, R.; Jeyaprakash, B.G. CeO_2 thin film as a low-temperature formaldehyde sensor in mixed vapour environment. *Bull. Mater. Sci.* **2014**, *37*, 1293–1299. [CrossRef]

34. Nagaraju, P.; Vijayakumar, Y.; Choudhary, R.J.; Ramana Reddy, M.V. Preparation and characterization of nanostructured Gd doped cerium oxide thin films by pulsed laser deposition for acetone sensor application. *Mater. Sci. Eng., B* **2017**, *226*, 99–106. [CrossRef]

 *biosensors*

Review

Evolution of Electronic Noses from Research Objects to Engineered Environmental Odour Monitoring Systems: A Review of Standardization Approaches

Domenico Cipriano [1] **and Laura Capelli** [2,*]

[1] Ricerca sul Sistema Energetico (RSE), via Rubattino 54, 20134 Milano, Italy; domenico.cipriano@rse-web.it
[2] Politecnico di Milano, Department of Chemistry, Materials, and Chemical Engineering "Giulio Natta", Piazza Leonardo da Vinci 32, 20133 Milano, Italy
* Correspondence: laura.capelli@polimi.it

Received: 17 April 2019; Accepted: 27 May 2019; Published: 31 May 2019

Abstract: Since electronic noses are used more and more for air quality monitoring purposes, and in some countries are starting to have a legal value, there is a need for standardization and programs for the quality verification of instruments. Such quality programs have the aim to guarantee the main characteristics of the instrument for both the final user and local authorities, let the user establish a suitable maintenance procedure and give information on measurement uncertainty. One critical aspect when dealing with electronic noses for environmental odour monitoring is that environmental odours are complex mixtures that are not repeatable nor reproducible, giving that they are not suitable for quality verifications. This paper aims to review and discuss the different approaches that can be adopted in order to perform quality checks on electronic noses (e-noses) used for environmental odour monitoring, thereby referring to existing technical standards, such as the Dutch NTA 9055:2012, the new German VDI 3518-3:2018, and the Italian UNI 1605848 project, which directly refer to electronic noses. Moreover, also the European technical standards that are prescriptive for automatic measuring systems (AMSs) are taken into consideration (i.e., EN 14181:2014 and EN 15267:2009), and their possible applicability to electronic noses is investigated. Finally, the pros and cons of the different approaches are presented and discussed in the conclusions section.

Keywords: air quality; technical standards; quality protocols; emission monitoring; sensor arrays; performance testing; minimum requirements

---

1. Introduction

Since the first report of an electronic nose based on chemical sensors and pattern recognition in 1982 [1], the instruments have experienced a significant development; they have been studied by several research groups over the world for a number of diverse possible applications in different sectors. The most interesting and promising sectors for the application of electronic noses that can be found in the scientific literature concern the food industry (e.g., [2–6]), medical diagnosis (e.g., [7–13]), and environmental monitoring (e.g., [14–16]).

It is a fact that, despite several promising results, practical applications of electronic noses in real-life are still limited [17]. Besides some technical criticalities, comprising for instance sensor lack of sensitivity/selectivity [18,19], interference with temperature and humidity [20], and drift [21–23], the lack of standardization also represents an important limit towards large-scale diffusion of electronic noses at an industrial level [15,24]. Given the wide range of different electronic noses available on the market, often based on different sensor types [25–27], data processing and pattern recognition systems [28,29], and/or functioning principles [4,30], it is very important that precise procedures for the verification of the instrument suitability for the desired application and its utilization are established [24].

Especially in the environmental field, in the last 15 years, electronic noses have become more and more popular air quality monitoring tools. As a matter of fact, they currently represent the only method available for the continuous monitoring of odours in the field [15,31].

Odour pollution is nowadays a serious environmental problem, and one of the major causes of citizens' complaints to local authorities. For this reason, odours are now subject to control and regulation in many countries [32,33], thus making it necessary to have reliable and accurate methods for the assessment of odour impacts. Indeed, dynamic olfactometry, which is the reference method for the measurement of odour concentration, is intrinsically discontinuous, and applies to the quantification of emission sources, as stated in the scope of the reference standard for dynamic olfactometry [34]. Dispersion models are commonly applied to evaluate how odour emissions disperse from the source to the receptors and to evaluate odour impacts [31,33]. The input parameter that defines odour emissions in dispersion models is the so called "odour emission rate" (OER), expressed in odour units per unit time (i.e., ou_E/s). However, there are many situations in which source characterization and the estimation of a representative OER may become extremely complex, for which the use of electronic noses is particularly indicated for odour impact or exposure assessment purposes. Such cases include for instance sources having variable emissions over time, whereby it is difficult to associate a specific OER to every hour of the simulation domain. Such variable emissions are typical of discontinuous productions, including for instance plants working "on order", which work just a few hours a day (e.g., asphalts production), or manufacture different products depending on customers' requests (e.g., pharmaceuticals). Another critical case is represented by diffuse sources, such as not-ventilated sheds or tanks, for which the estimation of the emitted air flow, which is necessary as input data for the dispersion model, is not always possible [35]. However, it is worth mentioning that, in recent years, great efforts are being made in the field of complex source characterization (e.g., [36,37]).

For this reason, in the field of environmental odour monitoring, electronic noses are rapidly turning from being only scientific and research objects to proper air quality monitoring tools. Besides examples describing applications of e-noses as air quality monitoring tools for odour impact assessment, alone or in combination with other techniques [38–40], there are also some situations in which e-noses are prescribed on a regulatory level [41,42]. It is clear that, when odour monitoring data produced by electronic noses start having a legal value, the need arises to have standards and quality programs allowing to ensure the quality of the whole monitoring process. As a general rule, standards play an important role for developing functional and reliable products for the global marketplace: they typically provide performance criteria that can be used to optimize the reliability and safety of new products [43]. According to this, standardized quality protocols are particularly needed for the instruments' performance verification.

As a general rule, such quality programs have the aim to guarantee the main characteristics of the instrument for both the final user and the local authorities, let the user establish a suitable maintenance procedure and give information on measurement uncertainty [44].Given the intrinsic complexity of electronic nose features, and the wide variety of instrument types—sometimes based on different functioning principles and sensor types [16,25,30]—available on the market, standardization in this field should not concern the instrument hardware, but requirements on its performance could be fixed. This approach allows the instrument to be considered as a "black box", by only taking into account the output metrics related to a given stimulus (input), thus ignoring the model that is used to transform the sensor signals into this output. One critical aspect regarding the application of electronic noses to environmental odour monitoring is that such odours are typically complex mixtures of hundreds of different compounds (e.g., [45–47]), and thus not repeatable nor reproducible. Environmental odours are therefore unsuitable for quality verifications, which require the standard reference materials that are tested to be repeatable and reproducible. However, some attempts of standardization have been carried out over the years, and will be discussed further on in this paper.

This paper has the aim of highlighting the importance of establishing suitable quality protocols applicable to electronic noses used as environmental odour monitoring systems and to describe some different approaches that can be adopted in order to perform quality checks on such instruments, thereby reviewing the relevant standards and publications in this field, and critically discuss the pros and cons related to their practical applicability.

For this purpose, the paper is divided into three parts.

The first part (Section 2) represents a state-of-the-art overview of how, in the last 15 years, electronic noses for environmental odour monitoring have evolved from a purely scientific and research level to become proper air quality monitoring tools used by plant operators as well as by local authorities for the management of odour issues through quantification of exposure or identification of emission sources. This state-of-the-art does not claim to provide an exhaustive review of all literature studies describing the application of electronic noses for environmental odour monitoring, nor to describe all currently available different electronic nose technologies, for which other extensive review papers already exist [4,15,16,48,49], but it is limited to those works proving the evolution of such instruments from research objects to regulatory tools.

The second part (Sections 3 and 4) reviews the relevant existing technical standards and guidelines that directly refer to electronic noses or that could possibly be applied to them. In more detail, Section 3 gives a short historical overview of the standardization attempts that have been made in Europe related to electronic noses, although not all necessarily referring exclusively to their environmental applications. Then, Section 4 briefly describes the relevant technical standards referring to other instruments intended for air quality monitoring, i.e., automatic measuring systems (AMSs). Although electronic noses are not AMSs, their implementation for emission and ambient air monitoring purposes, arises the question of the possibility to assimilate them with AMSs

Finally, the last part of this review (Section 5) represents a critical discussion of the possible approaches for the development and the application of standardized quality protocols for electronic noses intended specifically for environmental odour monitoring. In this section, possible alternative or complementary schemes for electronic noses qualification are proposed and discussed. As previously mentioned, such quality protocols are fundamental in order to properly characterize the instruments in terms of performance characteristics, measurement uncertainty, and effective applicability.

It is important to highlight that this paper only focuses on electronic noses used for monitoring odour as a whole, and not for the detection or identification of odorant substances.

2. State-of-the-Art of Electronic Nose Applications as Environmental Odour Monitoring Tools

The aim of this section is to give a brief overview of relevant examples of electronic nose applications for environmental odour monitoring, thereby focusing on those works that aimed to promote them as effective air quality monitoring tools. As previously mentioned, this section does not claim to give an exhaustive review of electronic nose applications for environmental odour monitoring, which have been already discussed in a recent review paper [15], but it has the aim only to describe their evolution in time from research objects to potential regulatory tools.

Historically, one of the first important works dealing with the use of an electronic nose for odour analysis and identification was reported by Nicolas et al. in 2000 [50]. This work describes the use of very simple instruments based on SnO_2 sensors for measurements around real sources of malodour in the environment such as compost facilities, waste water treatment plants, rendering plants, etc. giving promising classification results with discriminant analysis and principal component analysis. The authors also highlight the influence of atmospheric conditions on the sensor responses and the necessity to carry out repeated training over time in order to compensate sensor drift. In a more recent work [51], the same authors describe the application of five electronic noses, each comprising six Metal Oxide Semiconductor (MOS) sensors from Figaro®, for the assessment of odour annoyance near a compost facility. The study proves the system to be sufficiently efficient to predict in real time possible odour annoyance in the surroundings of the plant, even though the approach suffers from various uncertainties, from the sensors to the final determination of the distance of downwind annoyance.

One of the first examples in the literature in which the electronic nose is proposed as a methodology to quantitatively determine an odour impact was described by Sironi et al. in 2007 [52]. In this case, two electronic noses, each equipped with six MOS sensors, were used for monitoring odours from a composting plant. After training, one instrument was installed at the plant fence-line, while the second instrument was installed at a receptor located at about 4.3 km from the plant under investigation. The electronic noses analysed the air every 12 min for a 4-day period, then the sensor responses were analysed with the aim of classifying the analysed air into the different olfactory classes considered for training. This way, odour exposure could be assessed in terms of relative recognition frequency of odours from the monitored plant. This study also provides a sort of instrument performance evaluation by comparison of the outputs of the electronic nose installed at the receptor with the recordings of odour episodes of the residents. These data were presented in a confusion matrix and an accuracy index for classification was evaluated, which was equal to 72%. This result was considered as satisfactory, despite the reported influence of atmospheric humidity and sensor drift.

Another study in which the electronic nose responses were evaluated in combination with other sensorial observations (e.g., odour complaints reports and odour observations of experts) was reported by Milan et al. in 2012 [53]. This work describes a huge monitoring program aiming to map the odour impact in the Port of Rotterdam by using 40 fixed and four mobile electronic noses for a 3-year period. The objectives of investigating the electronic nose potential as an odour management tool for reducing odour exposure, as well as a safety management tool for the fast recognition of accidental gases resulting in incidents was considered as promising, although the validation procedure involving the comparison of instrumental and sensorial odour observations was not detailed in this work.

A different interesting application of electronic noses in this field was described by Chirmata et al. in 2015 [54], where electronic noses were used in order to implement a system for the continuous monitoring and tracking of odours in the city of Agadir. In this case, meteorological data were used in order to follow instantly the odour level in the study area and to identify the emission sources.

Finally, a very recent example of electronic nose implementation as an environmental odour monitoring tool in Italy is given by Licen et al. [55], who describes a 4-month survey close to a steel plant in Trieste. In that case, self-organizing maps proved to be a useful tool for visualizing the dynamic evolution of the system with time, thus allowing the identification of possible sources of malodour and evaluate frequency and duration of odour episodes.

Besides these examples, which describe the application of electronic noses as air quality monitoring tools for odour impact assessment, it is worth mentioning here that there are some cases in which the use of electronic noses for odour monitoring is foreseen on a regulatory level.

One first significant example is the French regulation about rendering plants [41], which, in article 46, foresees a reduction of the periodic odour measurement campaigns if a representative and permanent measurement is carried out by means of electronic noses.

Another very interesting example was recently presented by Cangialosi et al. in 2018 [42]. In this case an electronic nose was used in combination with an H_2S continuous analyser in order to provide a continuous measurement of the odour concentration at the fence-line of a sanitary landfill. An automatic alert to the local authority was set when the odour concentration measured by the electronic nose exceeded 500 ou_E/m^3 for more than 5 min or when two consecutive H_2S concentration measurements at 5 min intervals exceeded 20 ppb. Despite the very interesting application, the results of the study showed that the odour concentration values estimated by the electronic nose were poorly correlated with the H_2S concentration measurements.

It is worth highlighting that such prescriptions involving the installation of an electronic nose for continuous odour measurement around some plants (and specifically landfills) are now becoming a common trend in the Region of Puglia, in Southern Italy.

3. History of Standardization Attempts in the Field of Electronic Olfaction

This section gives a brief historical overview of the standardization attempts that have been made in Europe in the field of electronic olfaction. These do not necessarily refer only to electronic noses for environmental applications.

3.1. First Standardization Attempts: The Second Network on Artificial Olfactory Sensing (NOSE II)

As reported in a recent and very interesting opinion paper by T. Nagle and S. Schiffman [43], the first attempt for standardization in the field of electronic olfaction was carried out in 2001, under the sponsorship of the European Commission. The Second Network on Artificial Olfactory Sensing (NOSE II) [56] was constituted and its main task was to work out its own recommendations for standards and to foster their use in the sensor and e-nose community. The work was organized in three working groups (WGs), dealing with the following topics:

- standard data format for electronic nose data
- calibration and standardization
- hardware–software interfaces.

However, in the end, no specific standards were completed. The reasons given were: (i) the large number of available sensor types and electronic nose technologies; (ii) the WG goals were too broad; (iii) failure by the industry to establish a generic electronic nose technology; (iv) no broad industry support for a common data format; and (v) fragmented markets with different application requirements [43].

3.2. The NTA 9055.2012

After that, in 2012, the national standardization body of the The Netherlands (NEN) was the first one who succeeded in releasing a technical agreement document (NTA 9055:2012 [37]) regarding the specific use of electronic noses for the monitoring of odour emissions that may cause odour nuisance or safety risks.

As stated in the scope of this document: "the purpose of NTA 9055 is to draw up a list of requirements for using an electronic nose (e-nose) to detect changes in the composition of the ambient air".

The scope defines the following fields of application:

- Continuous monitoring: since dynamic olfactometry does not allow for continuous odour monitoring, electronic noses can be used for this purpose. It is claimed that "continuous monitoring using e-noses, combined with a knowledge of current process and weather conditions, makes it possible to identify the cause of odour nuisance in a targeted way".
- Information for risk assessment: electronic noses are proposed as tools enabling quick gathering of information concerning the dispersion of gaseous substances in the case of sudden major emissions, e.g., as a result of an incident. The document affirms that this information could possibly be used as a basis for a risk assessment.
- Emission detection and process monitoring: it is stated that the use of electronic noses for emission detection in industrial applications may help to optimise the process and minimise odour nuisance.

Then, after a brief general description of the electronic nose technology given in Section 1 (and normative references, definitions, and abbreviations reported in Sections 2–4), Section 5 aims to describe the methodology for using an electronic nose. First, the principles of electronic nose training are generically reported. Training involves the recording of the electronic nose signals when exposed to air with a given composition and the correlation of these data with the data acquired by sensory or instrumental analysis of the same air, so that "relevant sensory or instrumental perceptions can be reproduced if the same electronic nose data is recorded". It is specified that training can be carried out at the measurement site or in a laboratory. In both cases the procedure involves collecting the electronic nose data together with information provided by other means of detection and then correlating the electronic nose data with the other information.

Then, Section 5 of this document gives a general description of electronic nose networks (Section 5.2) and of mobile electronic noses (Section 5.3). Finally, Section 6 very briefly gives some indications about sampling, thereby referring to other existing norms.

Although this technical document has the undeniable value of having been the first technical document published by a standardization body regarding electronic noses, it has the drawback of being extremely synthetic and too generic to achieve the goal of standardization of the procedures for the application of these instruments in the environmental and safety sectors. Despite the statement of the scope of the document, no requirements for the use of electronic noses are defined, except for a generic description of the instrument training. What is totally missing for a standard is a description of the procedures for the verification of the electronic nose functionality nor the validation of the instrument outputs.

After the publication of this first technical document, with all its limitations, the need for standardization on this topic at a European level became evident. For this reason, a few years later, in 2014, the European Committee on Standardization (CEN) promoted the constitution of a working group dedicated to the draft of a standard on instrumental odour monitoring (WG41), whose activity will be detailed in the next section.

3.3. The CEN TC/264 WG41 "Instrumental Odour Monitoring"

As previously mentioned, after the publication of the NTA 9055:2012, a new European working group (WG41) was established within CEN TC/264 on air quality to draft a standard related to instrumental odour monitoring systems. The WG41 was composed by experts nominated by national standardization bodies from European countries including Belgium, France, Germany, Italy, The Netherlands, Spain, and the UK [58].

The scope of this standard is to specify requirements for instrumental odour monitoring systems (IOMS) for the monitoring of odour in ambient air and in emissions to ambient air. Indoor air was excluded from the scope of the WG. The primary application is to generate odour metrics that are relevant indicators for the presence and attributes of odour as would be perceived by human observers. A benefit of instrumental odour monitoring systems is that they can be used for continuous measurement [58].

The working item intentionally refers to IOMS and not to electronic noses, in order to include any generic "instrument" that can be applied to the monitoring of environmental odours, independently from its functioning principle or sensing technology. Odour is here considered as a whole, thereby referring to "odour" as the "sensation perceived by means of the olfactory organ in sniffing certain volatile substances", and not to single odorants. Moreover, given the wide range of different devices for instrumental odour monitoring available on the market, based on a variety of different functional principles for gas sensing, the technical design of such systems, their calibration, training and the mathematical model that is used to convert sensor signals into output metrics are not part of this standard, which considers the system as a "black box".

According to this approach, the work of the WG was focusing mainly on the validation regimes that can be used to prove performance claims. This included defining specific objectives and limitations, thus establishing procedures aiming to verify the instrument suitability for a specific application and its utilization, within defined boundary conditions.

The validation process consists of comparing the IOMS output metrics with odour assessment metrics obtained with suitable reference methods. The reference methods for odour measurement involve the use of human assessors.

For the task of the identification odour presence and odour classification the reference method is represented by field inspections using panel members according to EN 16841:2016 (part 1 or part 2) [59,60]. The EN 16841:2016 is a European Norm that standardizes the field inspection method for odour assessment in the field by means of qualified panel members. In more detail, part 1 of the standard describes a method ("grid method") for determining the level of odour exposure within a defined assessment area over a sufficiently long period of time to be representative for the meteorological conditions of that location. Part 2 describes a method ("plume method") for the determination of the extent of recognizable odours from a specific source using direct observation under specific meteorological conditions.

The reference method for validating the IOMS capability to quantify odours by providing an odour stimulus indicator value is dynamic olfactometry, as described by the EN 13725:2003 [34]. This standard defines the reference method for the determination of the odour concentration of an odorous gas sample using dynamic olfactometry with human assessors, thus providing a common basis for evaluation of odour emissions.

As already mentioned, up to now, the activity of the WG has been focusing more on the instrument final validation than its quality check. The main difficulties related to the development of a European Standard on this topic are associated with the fact that the currently existing technologies and applications in the different European countries, and therefore the expectations of the national representatives in the WG, are sometimes very different from each other.

However, since the activity of the WG is still ongoing and under discussion, it will be not described further in this paper.

3.4. The VDI 3518-3:2018

Very recently, in December 2018, the German VDI (Association of German Engineers) published a guideline relating specifically to odour measurements with electronic noses, i.e., the VDI 3518 Part 3 "Multigas sensors—Odour-related measurements with electronic noses and their testing" [61].

This guideline represents an important step forward towards standardization of electronic noses, although it does not refer specifically to the environmental monitoring of odours, but also to other fields of odour measurements. In more detail, the following application categories were identified:

- comfort
- diagnosis
- process monitoring
- safety.

One interesting aspect of the guideline is that it defines three different measurement tasks (functionalities) for electronic noses:

- differentiation, i.e., detecting the presence of odours;
- identification. i.e., determining the odour type;
- quantification, i.e., determining the odour intensity.

As stated in this guideline, since electronic noses can be used for a range of different odour-related measurements, the prerequisite is general suitability for the planned application.

For this reason, the VDI guideline describes a procedure for the instrument verification, which is based on three steps.

First, minimum equipment specifications for electronic noses have to be defined. For the formal testing of the equipment's suitability, data on usage, on the construction, on operating, and on the basic functionality have to be submitted.

As a second step, metrological functionality of the electronic nose has to be tested. Although demonstrating the metrological functionality of electronic noses is not a definitive proof of their suitability for the specific odour-related application, it has to be considered as a prerequisite for reliable operation of the instrument. If the metrological functionality testing is not passed, then it could hardly be expected that the instrument will be suitable for odour measurement in the desired final application. The test shall prove that, when exposed to a test gas, mechanical effects, electric interference, and climatic ambient factors, the electronic nose outcome shall not be affected by more than a deviation of 30% from the reference value.

Finally, once the metrological suitability of the electronic nose has been demonstrated, further testing is required in order to demonstrate the basic suitability of electronic noses for odour-related measurements. The tests can demonstrate the correlation between the electronic nose outputs and the results of an olfactory (sensorial) odour measurement carried out with human assessors.

Performance testing both for metrological functionality and for suitability for odour measurement has to be carried out with at least three standard test gases, containing odorants with known properties, to be chosen from a list provided in annex B of the guideline. Appropriate tests and odorants are to be chosen and the concrete test conditions agreed in accordance with the intended application.

The principle of the guideline is very interesting, since it tries to adopt a similar verification logic as those applied for other measurement instrumentation, but still considering the peculiarities of the electronic nose with respect to other chemical analysers. However, since the guideline is not application-specific, it does not account for the peculiarities of the single odour-related measurement. As a matter of fact, testing an electronic nose destined to environmental odour monitoring should entail the verification of different requirements as those that are needed, for instance, for medical diagnosis.

As previously mentioned, the guideline fixes minimum performances of the instruments to be guaranteed. The drawback of this approach is that the electronic nose technology is still very "young" and probably not mature enough to have fixed minimum requirements, which might, in the end, represent a limitation for further instrument development. Moreover, the minimum requirements fixed in this guideline are based on industrial certifications, thus hardly applicable to research prototypes and products of small manufacturers.

3.5. The Italian UNI1605848 Project

Very recently, in February 2019, the Italian Standardization Body UNI proposed a specific standard on the determination of odours by means of IOMS (instrumental odour monitoring systems) and their qualification. As already mentioned in Section 2, in Italy the use of electronic noses as air quality monitoring tools has grown significantly in the last few years, given that in some specific applications the electronic nose outputs have a legal value. As a consequence, this standard is the expression of an urgent need in Italy to provide the local authorities and the final users with an adequate normative text allowing for qualification of extremely different instruments that are proposed for environmental odour monitoring purposes.

The interesting and innovative thing about this standard is that it defines three quality assurance levels for electronic noses, following the principles of the EN 14181:2014 referred to automatic measuring systems (AMSs) for the continuous monitoring of emissions and ambient air, which will be described in the next section.

As a first step, electronic noses for environmental odour monitoring should undergo an initial metrological verification. This verification should be carried out by the manufacturer, in order to declare the instrument properties before its installation in the field. The parameters that shall be defined are, among others, the expected sensor responses to reference gases (whose composition is not specified and shall be chosen by the manufacturer) at zero and span, the effect of temperature and humidity on the sensor responses, and the response time T90. By definition, T90 is the time needed for a detector to measure 90% of the applied concentration level. In the case of electronic noses, T90 would be the time needed for the sensor to reach 90% of its maximum response to the applied reference gas.

It is also required that the capability of the instrument to provide correct results coherently with the type of determination (i.e., odour presence, odour class, or odour quantity) is verified in the laboratory before its installation in the field. The standard does not specify how this should be done in detail; the manufacturer shall provide all information related to the metrological verifications in a specific report that describes the methodology adopted.

The usefulness of this information is that, during its application in the field, the electronic nose functionality can be tested periodically by verifying that its properties declared in this initial metrological verification are still satisfied.

The essential verification step is carried out after installation in the field in order to verify the IOMS functionality as a whole, thereby including training, according to the reference method EN 13725:2003. The procedures for the verification depend on the type of determination requested (i.e., odour presence, odour class, or odour quantity). For any type of determination, the verification shall involve the comparison of at least 15 measurements provided by the electronic nose and 15 simultaneous measurements conducted in conformity with the EN 13725:2003, in the conditions that are considered to be most representative of the application. Measurements shall be carried out on at least four different days and be distributed in at least 6 h for each test day, in order to cover a 24-h cycle. The standard does not describe how the verification shall be carried out in detail, but it is required that a verification report is produced reporting all the relevant information about the test methods and results.

A mathematical approach is proposed for the evaluation of the deviation of the odour quantities provided by the electronic nose from the odour concentrations measured by dynamic olfactometry according to EN 13725:2003.

This standard, despite not being very detailed about the testing procedures, has the big advantage of being the only approach that is based on a multi-level verification of the electronic nose, which is fundamental in order to enable the instrument qualification in every phase of its life as an air quality monitoring tool.

4. Relevant Technical Norms Related to Other Instruments for Air Quality Monitoring

Besides the technical documents or standardization attempts described above, which are related directly to electronic noses, but not necessarily to air quality monitoring, other technical norms exist that refer to other instrumentation for air quality monitoring, which can be considered as inspiring for the development of quality protocols for electronic noses to be used in the field of environmental odour monitoring.

This is the case for the so called "automatic measuring systems" (AMSs), which are automatic analysers used for the continuous monitoring either of emissions or of ambient air.

Starting from 2000, CEN experts made a great effort to define all the aspects of AMSs. In more detail, the two documents that define the basic structure of AMSs are the standards EN 14181:2014 (first edition was released in 2005) and the EN 15267 series [44]. These technical standards, which describe the quality programs that must be followed in the realisation, validation and management of an AMS, will be briefly described in this section; then their applicability to electronic noses will be discussed in the next section, dedicated to the description of the approaches that can be adopted for the development of quality protocols for instrumental odour monitoring systems.

4.1. The EN 14181:2014

The Standard EN 14181:2014 [62] specifies the procedures for establishing quality assurance levels (QALs) for AMS installed on stationary sources for the determination of the flue gas components and other parameters. The following levels are defined [44].

- QAL1: the AMS, intended as the entire system—from the sampling up to the final result output—has to fulfil both general requirements (e.g., quality and safety requirements, availability, stability, sensitivity) as well as specific requirements related to the application (e.g., matrix of flue gas, interferents, emission limits, type of installation, weather conditions). For this purpose, a complete evaluation of the expected performance shall be carried out based on a detailed engineering project of the system, thereby using the mathematical formulations given by the Standard. The QAL1 process is considered complete when, based on the AMS design, it is possible to ensure that the uncertainty of the AMS is always below a given value.
- QAL2: QAL2 entails the initial technical verification of the system hardware and software after installation, in order to verify both the compliance with the design and especially the calibration of the system. QAL 2 verifications shall ensure that AMS measurements are reliable with the relevant

standard reference method (SRM). For every pollutant or chemical compound of interest, the relevant SRM represents the only "legally binding" method for the limit verification of that compound.

- QAL3: If QAL2 is accomplished, the AMS enters into "normal service". After that, the QAL3 procedure involves the establishment of a continuous Quality Assurance (QA)/Quality Control (QC) plan, in order to guarantee that the AMS is fully operative over time.
- Annual surveillance test (AST): every year, an independent verification test is required to check the AMS operation in order to verify QA/QC and maintenance procedure, or to solve the non-conformities eventually raised.

4.2. The EN 15267-1/2/3:2009

EN 15267-1/2/3:2009 [63–65] specifies the general principles, including common procedures and requirements, for the product certification of AMSs for monitoring ambient air quality and emissions from stationary sources. This product certification consists of the following sequential stages [44]:

- performance testing of an automated measuring system;
- initial assessment of the AMS manufacturer's quality management system;
- certification;
- surveillance.

In more detail, the scope of EN 15267 Standards series is:

- The specification of requirements for the manufacturer's quality management system, the initial assessment of the manufacturer's production control and the continuing surveillance of the effect of subsequent design changes on the performance of the AMS. It also serves as a reference document for auditing the manufacturer's quality management system.
- The definition of the performance criteria and test procedures for the AMS. It provides the detailed procedures covering the QAL1 requirements of the EN 14181:2014 and, where required, the input data to be used in QAL3.

This European Standard applies to the certification of all AMS for monitoring ambient air quality and emissions from stationary sources for which performance criteria and test procedures are available as European Standard.

5. Approaches for the Development of Quality Protocols for Instrumental Odour Monitoring Systems

For the specific application of electronic noses to environmental odour monitoring, different approaches could possibly be adopted for the development of procedures for the instrument qualification. These approaches are discussed in this section.

5.1. Approaches for the Qualification of Electronic Noses Described in the Scientific Literature

Other more or less similar approaches for e-nose performance testing have been proposed in the scientific literature. An example of approach that can be used for qualification of electronic noses to be used as environmental odour monitoring systems is given in References [24,66].

In more detail, the first example [24] proposes a testing procedure for performance evaluation aimed at the definition of minimum performance requirements referred to electronic noses to be used for environmental odour monitoring at receptors.

In this study, the following aspects were deemed important for qualifying an electronic nose for environmental odour monitoring:

- the capability of giving repeatable and reliable responses under variable atmospheric conditions; indeed, variations of temperature and humidity are particularly critical for e-nose environmental outdoor use ("invariability of responses to variable atmospheric conditions");

- the sensitivity to odours: if e-noses are used at receptors they are likely to be exposed to very diluted concentrations, for this reason, instruments shall have a very high sensitivity (i.e., a low "detection limit");
- the capability of correctly classifying the detected odours, by recognizing their provenance and attributing them to the correct olfactory class ("classification accuracy").

Then the study describes as example of detailed procedure adopted to test a commercial electronic nose towards those aspects. According to the proposed procedure, the abovementioned aspects shall be tested using standard test gases. The use of standard test gases is preferred over "real" environmental samples, because of the intention of guaranteeing repeatability and reproducibility of the procedure, in order to make it possible to compare the performances of different instruments tested in different labs or at different times and conditions.

The second example [66] proposes a procedure for electronic nose performance testing based on a similar approach. In this case, the instrument performances are evaluated through the definition of an experimental protocol, which is structured into different levels.

The paper focuses on the first two levels of testing, which involve specific tests with standard test gases, including for instance chemical compounds that are representative of common environmental odours. These specific tests allow for the evaluation of the electronic nose limit of detection, lower detection limit towards those target compounds and repeatability of responses to a given stimulus. This type of testing is not related to specific applications and therefore raw signals are considered instead of instrument output for performance evaluations. On the other hand, the third level of testing is related to the specific application, and thus the performance of the trained system is evaluated in the field in terms of classification accuracy. This third level of testing was not described in the paper [66].

The two abovementioned papers have in common that the proposed approach is based on the idea of verifying some fundamental aspects (e.g., lower detection limit, response repeatability, capability of compensating humidity and temperature variations, and capability to classify different odour types correctly) by performing performance testing using standard test gases.

It is important to highlight that since standard test gases are never fully representative of what happens in "real" environmental conditions, these tests shall not be considered as a sufficient condition to prove the electronic nose suitability for the specific application. What is described here is a sort of pre-qualification, which always needs to be followed by a validation in the field.

This is not very dissimilar from the approach proposed in the VDI 3518-3:2018; the main difference is that here no minimum performance requirements are fixed.

5.2. Applicability of EN 14181:2014 and EN 15267:2009 to Electronic Noses

5.2.1. Electronic Nose as AMS?

As described in the previous section, the Technical Standards EN 14181:2014 and EN 15267:2009 define the quality protocols that must be applied to AMS, which in facts are systems for the continuous sampling, measurement and control of pollutants in emissions and ambient air.

In a very recent paper by Cipriano [44], the author proposed the possibility of implementing the features of the technical standards for AMSs to electronic noses, thereby focusing on qualification and maintenance, and on the uncertainty aspects.

Although electronic noses are not AMSs, and thus this proposal may sound as a provocation, the implementation of electronic noses for emission and ambient air monitoring purposes, as described in the previous section, the question arises of the opportunity to partially assimilate them to AMSs. A possible integration of such instruments into the universe of AMSs could be advantageous in order to define the uncertainties associated with their use as air quality monitoring tools.

Indeed, an AMS device is intended to be used for continuous legal use, so it must ensure reliability, integrity and data security. It shall allow the calculation of both the uncertainty on the measured values vs. a SRM, and an independent verification of its metrological capabilities.

The possibility to apply the principles of the standards for AMSs to the verification of these same aspects is discussed here, as proposed in Reference [44].

5.2.2. Reliability, Integrity and Data Security

In order to allow QAL1 and QAL2 evaluation, electronic noses shall declare the performances of the entire system (sensors, data acquisition, processing, interfaces, etc.) and permit their verification. Also, in order to allow QAL3 procedures, it is necessary to have specific hardware and software solutions to implement periodic checks of the sensors array (Figure 1).

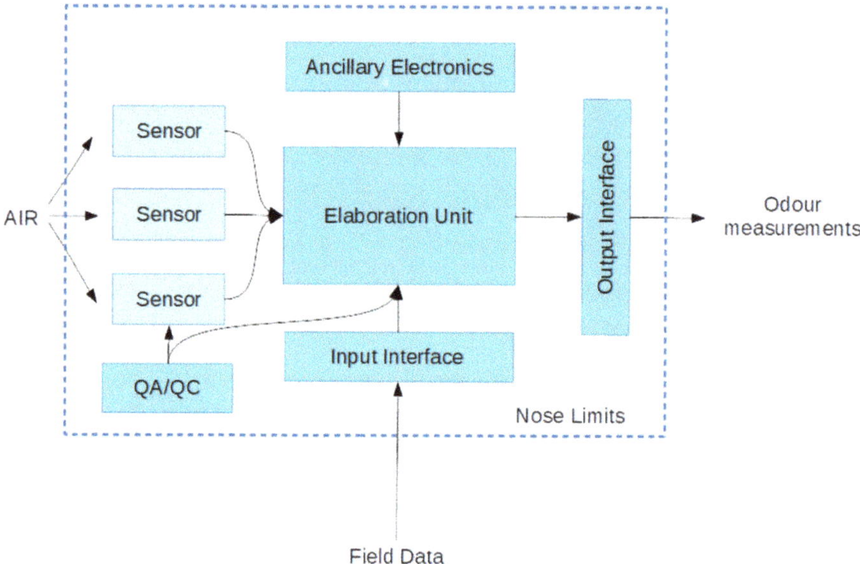

Figure 1. Electronic nose with quality assurance (QA)/QC control of the sensor array.

One of the biggest advantages of electronic noses is their flexibility due to possible modification of the software, by using different algorithms and calibration data. Such flexibility can be very useful for obtaining a good QAL2, so that the instrument can be trained to give results very close to the calibration values. However, this is also one of the biggest problems during ongoing verifications, as every modification done after QAL2 process implies the formal invalidation of its performances. E-nose structure shall ensure that, when the calibration process is finished, all the relevant data and configurations are locked and encrypted in order to prevent performances changes and unauthorized modifications.

5.2.3. Uncertainty

Nowadays all emission and ambient air monitors shall declare their "uncertainty budget" evaluated following principles of the ISO IEC Guide 98-3:2008 "Guide to expression of uncertainty in measurement" [67]. In this guide, the output of the measuring process is described by Equation (1) [44]:

$$y = f(x_1, x_2, \ldots, x_n). \tag{1}$$

The estimation of the total uncertainty $u(y)$ can be obtained by the propagation of the single uncertainty terms for each x_i, $u(x_i)$, by the means of Equation (2):

$$u(y) = \sqrt{\sum C_i^2 u^2(x_i)}, \tag{2}$$

where C_i is the sensitivity coefficient of the single x_i term, and $u(x_i)$ its uncertainty.

Function f shall cover all the measuring process, from the single sensor acquisition up to the final output, including calibrations, interferences, nonlinearities and software-related errors. For that, the use of a new CEN standard, the FprCEN/TS 17198:2017 "Stationary source emissions—Predictive Emission Monitoring Systems (PEMS)—Applicability, execution and quality assurance" could be useful. Such a standard is designed for emission prediction models, but could be easily adapted to electronic noses. Furthermore, its scope is to achieve conformity to EN 14181 and EN 15267 of software predictive systems and furnishes a simplified formula for uncertainty evaluation [44]:

$$u(y) = \sqrt{\left(u^2_{model} + u^2_{input} + u^2_{other}\right)}, \tag{3}$$

where u_{model} is the uncertainty of the mathematical model, u_{input} the uncertainty from sensors array, u_{other} the uncertainty due to parameters not included in the model, evaluated by confrontation with independent odour measurements used to calibrate the electronic nose.

5.2.4. Independent Verifications

An e-nose shall make it possible for the final user to verify its metrological capabilities, in order to validate its operational status and guarantee its calibrations are still aligned with QAL2 results. This means that, for instance, a specific gas matrix should be realised and used for periodic system validation.

6. Discussion and Conclusions

As stated in the introduction, this paper had the primary aim of discussing the need for the definition of quality protocols for electronic noses that are intended as air quality monitoring tools for the detection and the measurement of environmental odours. As a part of this discussion, this work also described some of the possible approaches that can be adopted for a standardization and qualification of such instruments, thereby reviewing the relevant documents—technical standards and scientific publications—on the issue.

In order to allow a better evaluation of these qualification approaches in terms of practical applicability to e-noses for environmental use, as well as the possible advantages and drawbacks related to their application, Table 1 schematizes the most relevant aspects relevant to the different approaches, together with their pros and cons.

Based on the information reported in Table 1, it is possible to appreciate the pros and cons of the different approaches. Leaving out the Dutch NTA 9055:2012, which is so generic that no real standardization/qualification approach is proposed, and the CEN TC/264 WG41 activity, which is still in progress and at a very preliminary stage, it is possible to make some more specific considerations about the other technical documents reviewed here.

The qualification approach based on the combination of the EN 14181:2014 and the EN 15267:2009 has the enormous advantage, compared to the others, that it is the only one validated. However, despite being a consolidated and effective approach, it is only valid (and validated) for monitoring systems that are not e-noses, and thus would need to be deeply re-adapted in order to make it applicable to odour measurement systems.

On the other hand, the other approaches, which are specific for e-noses and for odour measurement, to the best of our knowledge, are not validated yet. Therefore, up to now, there is no available data comparing how different instruments would perform by application of any of these testing protocols. The only data available in literature concern testing procedures that are not part of a standard [66,67].

Table 1. Schematization of the existing approaches that can be applied for e-nose qualification and standardization in environmental odour monitoring applications.

Approach	Nationality	Specific for E-Nose?	Specific for Environmental Use?	Principle of Standardization	Pros	Cons
NTA 9055:2012	The Netherlands	✓	✓	None	• Historical value: it is the first national technical document on e-noses	• Too generic • Specific testing procedures are not defined • No standardization approach is proposed
EN14181:2014 + EN15267:2009	Europe	✗	✓	Performance declaration	• Complete and validated approach (for AMS) • Stepwise approach with three different testing levels for performance verification • Allows comparison of systems based on different functioning principles	• Not referred to e-noses • Would need to be revised to make it applicable to odour measurement systems
VDI 3518-3:2018	Germany	✓	✗	Minimum requirements	• Very complete and detailed guideline • Specific for different e-nose applications • Identification of three different tasks (i.e., differentiation, identification and quantification), each involving specific testing procedures • Two-levels testing: (1) metrological functionality, and (2) suitability for odour measurement	• Measurement uncertainty is mentioned, but a method for evaluation of this uncertainty is not specified • Minimum requirements are fixed, and are mostly based on industrial certifications: this might be premature for a technology that is still under development, and consequently limit the development of research prototypes

Table 1. *Cont.*

Approach	Nationality	Specific for E-Nose?	Specific for Environmental Use?	Principle of Standardization	Pros	Cons
CEN TC/264 WG41	Europe	✓	✓	Not defined yet	• Standardization attempt on an international level: the WG includes experts from several European countries • Identification of three different tasks (i.e., differentiation, identification and quantification), each involving specific testing procedures • "Black-box" approach	• Still in progress, at a very preliminary stage • Not a stepwise approach: only focused on final validation • Does not allow comparison of different systems prior to their installation in the field • Not validated
UNI 1605848	Italy	✓	✓	Performance declaration	• Stepwise testing procedure inspired to the QALs of the EN 14181:2014 for AMS • Three types of possible determinations are defined (i.e., odour presence, odour class, or odour quantity), each involving specific qualification procedures • A method for the measurement uncertainty evaluation is proposed • Flexible approach: testing procedures are not fixed, but they shall be detailed together with the achieved performances in a specific report • This approach leaves some freedom in the characterization of instruments and in the development of new technologies	• Definition of the instrument pre-qualification testing procedures are lacking • The procedures for comparison with reference methods (e.g., EN 13725:2003) are not defined in detail for each type of determination • Not validated yet

One aspect that emerges from the information reported in Table 1, is that two different concepts for standardization/qualification can be distinguished. The first one involves the definition of minimum performance requirements. Even though the establishment of minimum requirements may result in a higher level of guarantee for the end user, this may entail the drawback of being a bit premature for a technology that is still under development. This is particularly true in consideration that e-noses for environmental use can be applied in a variety of different situations and with different purposes, for which different instrumental performance characteristics may be required. For this reason, it might be better to have a more flexible approach involving instrument testing and performance declaration without the necessity to achieve a minimum performance level.

Thus, a possible alternative approach for a quality protocol regarding electronic noses to be used for environmental odour monitoring would be a combination of the concept of performance testing with the requirements of the quality standards for AMS. This would entail a method for performance testing and ongoing quality procedures (QAL3), based on the principle that the instrument manufacturer shall specify the electronic nose characteristics, metrological functionality, and design limits, without fixing minimum performances. Thus, it is the final user who evaluates the instrument suitability and makes the choice of the most appropriate electronic nose features according to his specific needs, or, for instance, to the current regulatory requirements.

This is basically the approach followed by the Italian UNI1605848 project, which, despite its lack of details regarding the testing procedure, is the only multi-level verification approach allowing for the characterization of the different aspects of the electronic nose functioning in every phase of the instrument's life and application.

This approach seems to guarantee higher flexibility, thus making it applicable also to prototypal solutions, which is an important aspect for a technology that is still at an initial stage as proper air quality monitoring tool and in continuous development.

As a conclusion, application of specific quality programs to electronic noses for air quality monitoring is a delicate and complicated process, not yet deeply explored, and where there are continuous implementations of both the instruments and the relevant technical standardisation. However, the current status of the technology seems to be sufficiently mature to undergo such a process, which is necessary in order to make electronic noses a widespread and effective environmental odour impact assessment tool.

Funding: This research received no external funding.

Conflicts of Interest: The authors declare no conflict of interest.

References

1. Persaud, K.; Dodd, G. Analysis of discrimination mechanisms in the mammalian olfactory system using a model nose. *Nature* **1982**, *299*, 352–355. [CrossRef] [PubMed]
2. Di Rosa, A.R.; Leone, F.; Cheli, F.; Chiofalo, V. Fusion of electronic nose, electronic tongue and computer vision for animal source food authentication and quality assessment—A review. *J. Food Eng.* **2017**, *210*, 62–75. [CrossRef]
3. Gliszczyńska-Świgło, A.; Chmielewski, J. Electronic nose as a tool for monitoring the authenticity of food. A review. *Food Anal. Methods* **2017**, *10*, 1800–1816. [CrossRef]
4. Loutfi, A.; Coradeschi, S.; Mani, G.K.; Shankar, P.; Rayappan, J.B.B. Electronic noses for food quality: A review. *J. Food Eng.* **2015**, *144*, 103–111. [CrossRef]
5. Peris, M.; Escuder-Gilabert, L. A 21st century technique for food control: Electronic noses. *Anal. Chim. Acta* **2009**, *638*, 1–15. [CrossRef]
6. Ampuero, S.; Bosset, J.O. The electronic nose applied to dairy products: A review. *Sens. Actuators B Chem.* **2003**, *94*, 1–12. [CrossRef]
7. Capelli, L.; Taverna, G.; Bellini, A.; Eusebio, L.; Buffi, N.; Lazzeri, M.; Tidu, L.; Grizzi, F.; Sardella, P.; Latorre, G.; et al. Application and uses of electronic noses for clinical diagnosis on urine samples: A review. *Sensors* **2016**, *16*, 1708. [CrossRef] [PubMed]

8. Krilaviciute, A.; Heiss, J.A.; Leja, M.; Kupcinskas, J.; Haick, H.; Brenner, H. Detection of cancer through exhaled breath: A systematic review. *Oncotarget* **2015**, *6*, 38643. [CrossRef]

9. Wilson, A. Advances in electronic-nose technologies for the detection of volatile biomarker metabolites in the human breath. *Metabolites* **2015**, *5*, 140–163. [CrossRef]

10. Scarlata, S.; Pennazza, G.; Santonico, M.; Pedone, C.; Antonelli Incalzi, R. Exhaled breath analysis by electronic nose in respiratory diseases. *Expert Rev. Mol. Diagn.* **2015**, *15*, 933–956. [CrossRef] [PubMed]

11. Chen, S.; Wang, Y.; Choi, S. Applications and technology of electronic nose for clinical diagnosis. *Open J. Appl. Biosens.* **2013**, *2*, 39. [CrossRef]

12. D'Amico, A.; Di Natale, C.; Falconi, C.; Martinelli, E.; Paolesse, R.; Pennazza, G.; Santonico, M.; Sterk, P.J. Detection and identification of cancers by the electronic nose. *Expert Opin. Med. Diagn.* **2012**, *6*, 175–185. [CrossRef] [PubMed]

13. Wilson, A.D.; Baietto, M. Advances in Electronic Nose Technologies Developed for Biomedical Applications. *Sensors* **2011**, *11*, 1105–1176. [CrossRef] [PubMed]

14. Deshmukh, S.; Bandyopadhyay, R.; Bhattacharyya, N.; Pandey, R.A.; Jana, A. Application of electronic nose for industrial odors and gaseous emissions measurement and monitoring—An overview. *Talanta* **2015**, *144*, 329–340. [CrossRef] [PubMed]

15. Capelli, L.; Sironi, S.; Del Rosso, R. Electronic Noses for Environmental Monitoring Applications. *Sensors* **2014**, *14*, 19979–20007. [CrossRef]

16. Wilson, A.D. Review of electronic-nose technologies and algorithms to detect hazardous chemicals in the environment. *Procedia Technol.* **2012**, *1*, 453–463. [CrossRef]

17. Boeker, P. On 'Electronic Nose' methodology. *Sens. Actuators B Chem.* **2014**, *204*, 2–17. [CrossRef]

18. Nicolas, J.; Romain, A.C. Establishing the limit of detection and the resolution limits of odorous sources in the environment for an array of metal oxide gas sensors. *Sens. Actuators B Chem.* **2004**, *99*, 384–392. [CrossRef]

19. Bochenkov, V.; Sergee, V.G. Sensitivity, selectivity, and stability of gas-sensitive metal-oxide nanostructures. In *Metal Oxide Nanostructures and Their Applications*; Umar, A., Hahn, Y.B., Eds.; American Scientific Publishers: Valencia, CA, USA, 2010; Volume 3, pp. 31–52.

20. Mumyakmaz, B.; Ozmen, A.; Ebeoglu, M.A.; Tasaltin, C.; Gurol, I. A study on the development of a compensation method for humidity effect in QCM sensor responses. *Sens. Actuators B Chem.* **2010**, *147*, 277–282. [CrossRef]

21. Ziyatdinov, A.; Marco, S.; Chaudry, A.; Persaud, K.; Caminal, P.; Perera, A. Drift compensation of gas sensor array data by common principal component analysis. *Sens. Actuators B Chem.* **2010**, *146*, 460–465. [CrossRef]

22. Vergara, A.; Vembu, S.; Ayhan, T.; Ryan, M.A.; Homer, M.L.; Huerta, R. Chemical gas sensor drift compensation using classifier ensembles. *Sens. Actuators B Chem.* **2012**, *166–167*, 320–329. [CrossRef]

23. Fonollosa, J.; Fernandez, L.; Gutiérrez-Gálvez, A.; Huerta, R.; Marco, S. Calibration transfer and drift counteraction in chemical sensor arrays using Direct Standardization. *Sens. Actuators B Chem.* **2016**, *236*, 1044–1053. [CrossRef]

24. Eusebio, L.; Capelli, L.; Sironi, S. Electronic Nose Testing procedure for the Definition of Minimum Performance Requirements for Environmental Odor Monitoring. *Sensors* **2016**, *16*, 1548. [CrossRef] [PubMed]

25. Zohora, S.E.; Khan, A.M.; Hundewale, N. Chemical sensors employed in electronic noses: A review. In *Advances in Computing and Information Technology*; Springer: Berlin/Heidelberg, Germany, 2013; pp. 177–184.

26. James, D.; Scott, S.M.; Ali, Z.; O'hare, W.T. Chemical sensors for electronic nose systems. *Microchim. Acta* **2005**, *149*, 1–17. [CrossRef]

27. Arshak, K.; Moore, E.; Lyons, G.M.; Harris, J.; Clifford, S. A review of gas sensors employed in electronic nose applications. *Sens. Rev.* **2004**, *24*, 181–198. [CrossRef]

28. Yan, J.; Guo, X.; Duan, S.; Jia, P.; Wang, L.; Peng, C.; Zhang, S. Electronic nose feature extraction methods: A review. *Sensors* **2015**, *15*, 27804–27831. [CrossRef] [PubMed]

29. Scott, S.M.; James, D.; Ali, Z. Data analysis for electronic nose systems. *Microchim. Acta* **2006**, *156*, 183–207. [CrossRef]

30. Wilson, A.D.; Baietto, M. Applications and Advances in Electronic-Nose Technologies. *Sensors* **2009**, *9*, 5099–5148. [CrossRef] [PubMed]

31. Capelli, L.; Sironi, S.; del Rosso, R.; Guillot, J.M. Measuring odours in the environment vs. dispersion modelling: A review. *Atmos. Environ.* **2013**, *79*, 731–743. [CrossRef]

32. Nicell, J.A. Assessment and regulation of odour impacts. *Atmos. Environ.* **2009**, *43*, 196–206. [CrossRef]

33. Brancher, M.; Griffiths, K.D.; Franco, D.; de Melo Lisboa, H. A review of odour impact criteria in selected countries around the world. *Chemosphere* **2017**, *168*, 1531–1570. [CrossRef] [PubMed]

34. CEN. *EN 13725:2003. Air Quality—Determination of Odour Concentration by Dynamic Olfactometry*; Comitée Européen de Normalisation: Brussels, Belgium, 2007.

35. Capelli, L.; Dentoni, L.; Sironi, S.; Del Rosso, R. The need for electronic noses for environmental odour exposure assessment. *Water Sci. Technol.* **2014**, *69*, 135–141. [CrossRef]

36. Prata, A.A., Jr.; Lucernoni, F.; Santos, J.M.; Capelli, L.; Sironi, S.; Le-Minh, N.; Stuetz, R.M. Mass transfer inside a flux hood for the sampling of gaseous emissions from liquid surfaces—Experimental assessment and emission rate rescaling. *Atmos. Environ.* **2018**, *179*, 227–238. [CrossRef]

37. Invernizzi, M.; Ilare, J.; Capelli, L.; Sironi, S. Proposal of a method for evaluating odour emissions from refinery storage tanks. *Chem. Eng. Trans.* **2018**, *68*, 49–54.

38. Giungato, P.; de Gennaro, G.; Barbieri, P.; Briguglio, S.; Amodio, M.; de Gennaro, L.; Lasigna, F. Improving recognition of odors in a waste management plant by using electronic noses with different technologies, gas chromatography–mass spectrometry/olfactometry and dynamic olfactometry. *J. Clean. Prod.* **2016**, *133*, 1395–1402. [CrossRef]

39. Deshmukh, S.; Jana, A.; Bhattacharyya, N.; Bandyopadhyay, R.; Pandey, R.A. Quantitative determination of pulp and paper industry emissions and associated odor intensity in methyl mercaptan equivalent using electronic nose. *Atmos. Environ.* **2014**, *82*, 401–409. [CrossRef]

40. Stuetz, R.M.; Fenner, R.A.; Engin, G. Assessment of odours from sewage treatment works by an electronic nose, H_2S analysis and olfactometry. *Water Res.* **1999**, *33*, 453–461. [CrossRef]

41. JORF. Arrêté du 12 février 2003 relatif aux prescriptions applicables aux installations classes soumises à autorisation sous la rubrique 2730 traitement de sous-produits d'origine animale, y compris débris, issues et cadavres, à l'exclusion des activités visées par d'autres rubriques de la nomenclature, des établissements de diagnostic, de recherche et d'enseignement. Available online: https://www.legifrance.gouv.fr/affichTexte.do?cidTexte=JORFTEXT000000419523 (accessed on 31 May 2019).

42. Cangialosi, F.; Intini, G.; Colucci, D. On line monitoring of odour nuisance at a sanitary landfill for non-hazardous waste. *Chem. Eng. Trans.* **2018**, *68*, 127–132.

43. Nagle, H.T.; Schiffman, S.S. Electronic Taste and Smell: The Case for Performance Standards [Point of View]. *Proc. IEEE* **2018**, *106*, 1471–1478. [CrossRef]

44. Cipriano, D. Application of EN14181 and EN15267 to electronic noses: Challenge or provocation? *Chem. Eng. Trans.* **2018**, *68*, 139–144.

45. Parker, D.B.; Gilley, J.; Woodbury, B.; Kim, K.H.; Galvin, G.; Bartelt-Hunt, S.L.; Li, X.; Snow, D.D. Odorous VOC emission following land application of swine manure slurry. *Atmos. Environ.* **2013**, *66*, 91–100. [CrossRef]

46. Fang, J.J.; Yang, N.; Cen, D.Y.; Shao, L.M.; He, P.J. Odor compounds from different sources of landfill: Characterization and source identification. *Waste Manag.* **2012**, *32*, 1401–1410. [CrossRef]

47. Kim, K.H.; Jeon, E.C.; Choi, Y.J.; Koo, Y.S. The emission characteristics and the related malodor intensities of gaseous reduced sulfur compounds (RSC) in a large industrial complex. *Atmos. Environ.* **2006**, *40*, 4478–4490. [CrossRef]

48. Guillot, J.M. E-noses: Actual limitations and perspectives for environmental odour analysis. *Chem. Eng. Trans.* **2016**, *54*, 223–228.

49. Gebicki, J. Application of electrochemical sensors and sensor matrixes for measurement of odorous chemical compounds. *TrAC Trends Anal. Chem.* **2016**, *77*, 1–13. [CrossRef]

50. Nicolas, J.; Romain, A.C.; Wiertz, V.; Maternova, J.; Andre, P. Using the classification model of an electronic nose to assign unknown malodours to environmental sources and to monitor them continuously. *Sens. Actuators B Chem.* **2000**, *69*, 366–371. [CrossRef]

51. Nicolas, J.; Cerisier, C.; Delva, J.; Romain, A.C. Potential of a network of Electronic Noses to assess the odour annoyance in the environment of a compost facility. *Chem. Eng. Trans.* **2012**, *30*, 133–138.

52. Sironi, S.; Capelli, L.; Centola, P.; del Rosso, R.; Il Grande, M. Continuous monitoring of odours from a composting plant using electronic noses. *Waste Manag.* **2007**, *27*, 389–397. [CrossRef]

53. Milan, B.; Bootsma, S.; Bilsen, I. Advances in odour monitoring with E-Noses in the Port of Rotterdam. *Chem. Eng. Trans.* **2012**, *30*, 145–150.

54. Chirmata, A.; Ichou, I.A.; Page, T. A continuous electronic nose odor monitoring system in the city of Agadir Morocco. *J. Environ. Prot.* **2015**, *6*, 54–63. [CrossRef]

55. Licen, S.; Barbieri, G.; Fabbris, A.; Briguglio, S.C.; Pillon, A.; Stel, F.; Barbieri, P. Odor Control Map: Self Organizing Map built from electronic nose signals and integrated by different instrumental and sensorial data to obtain an assessment tool for real environmental scenarios. *Sens. Actuators B Chem.* **2018**, *263*, 476–485. [CrossRef]

56. Holmberg, M.; Eriksson, M.; Krantz-Rülcker, C.; Artursson, T.; Winquist, F.; Lloyd-Spetz, A.; Lundström, I. 2nd workshop of the second network on artificial olfactory sensing (NOSE II). *Sens. Actuators B Chem.* **2004**, *101*, 213–223. [CrossRef]

57. NEN. *NTA 9055:2012 Air Quality—Electronic air monitoring- Odour (nuisance) and safety*; Nederlands Normalisatie-instituut: Delft, The Netherlands, 2012.

58. Romain, A.C.; Capelli, L.; Guillot, J.M. Instrumental odour monitoring: Actions for a new European Standard. In Proceedings of the 2017 ISOCS/IEEE International Symposium on Olfaction and Electronic Nose (ISOEN 2017), Montreal, QC, Canada, 28–31 May 2017.

59. CEN. *EN 16841-1:2016 Ambient air—Determination of Odour in Ambient Air by Using Field Inspection—Part 1: Grid Method*; Comitée Européen de Normalisation: Brussels, Belgium, 2016.

60. CEN. *EN 16841-2:2016 Ambient Air—Determination of Odour in Ambient Air by Using Field Inspection—Part 2: Plume Method*; Comitée Européen de Normalisation: Brussels, Belgium, 2016.

61. VDI. *VDI 3518-3:2018. Multigas Sensors—Odour Related Measurements with Electronic Noses and their Testing*; Beuth Verlag: Berlin, Germany, 2018.

62. CEN. *EN 14181:2014 Stationary Source Emissions—Quality Assurance of Automated Measuring Systems*; Comitée Européen de Normalisation: Brussels, Belgium, 2014.

63. CEN. *EN 15267-1:2009 Air Quality—Certification of Automated Measuring Systems—Part 1: General Principles*; Comitée Européen de Normalisation: Brussels, Belgium, 2009.

64. CEN. *EN 15267-2:2009 Air Quality—Certification of Automated Measuring Systems—Part 2: Initial Assessment of the AMS Manufacturer's Quality Management System and Post Certification Surveillance for the Manufacturing Process*; Comitée Européen de Normalisation: Brussels, Belgium, 2009.

65. CEN. *EN 15267-3:2007 Air Quality—Certification of Automated Measuring Systems—Part 3: Performance Criteria and Test Procedures for Automated Measuring Systems for Monitoring Emissions from Stationary Sources*; Comitée Européen de Normalisation: Brussels, Belgium, 2009.

66. Li Voti, M.; Bax, C.; Marzocchi, M.; Sironi, S.; Capelli, L. Testing procedure for performance evaluation of electronic noses for environmental odour monitoring. *Chem. Eng. Trans.* **2018**, *68*, 277–282.

67. ISO. *ISO/IEC Guide 98-3:2008. Guide to the Expression of Uncertainty in Measurement (GUM:1995)*; ISO: Geneva, Switzerland, 2008.

Review

Use of Electronic Noses for Diagnosis of Digestive and Respiratory Diseases through the Breath

Carlos Sánchez [1,2], J. Pedro Santos [1] and Jesús Lozano [3,*]

[1] Institute of Physics Technology and Information (CSIC), 28006 Madrid, Spain; carlos@updevices.com (C.S.); jp.santos@csic.es (J.P.S.)
[2] Up Devices and Technologies, 28021 Madrid, Spain
[3] Industrial Engineering School, University of Extremadura, 06006 Badajoz, Spain
* Correspondence: jesuslozano@unex.es; Tel.: +34-924-289-300

Received: 12 January 2019; Accepted: 21 February 2019; Published: 28 February 2019

Abstract: The increased occurrence of chronic diseases related to lifestyle or environmental conditions may have a detrimental effect on long-term health if not diagnosed and controlled in time. For this reason, it is important to develop new noninvasive early diagnosis equipment that allows improvement of the current diagnostic methods. This, in turn, has led to an exponential development of technology applied to the medical sector, such as the electronic nose. In addition, the appearance of this type of technology has allowed the possibility of studying diseases from another point of view, such as through breath analysis. This paper presents a bibliographic review of past and recent studies, selecting those investigations in which a patient population was studied with electronic nose technology, in order to identify potential applications of this technology in the detection of respiratory and digestive diseases through the analysis of volatile organic compounds present in the breath.

Keywords: electronic nose; gas sensors; biomarkers; diseases; digestive system; respiratory system; volatile organic compounds; breath

1. Introduction

The relationship between aromas present in the breath and disease has been known by doctors for a several hundred years. The use of aromas to identify diseases dates back to the fourth century, when the doctor, based on experience, could determine what disease a patient was suffering. For example, a fruity aroma in the breath was identified as a sign of ketoacidosis associated with diabetes, and the smell of ammonia can be indicative of kidney failure. This method was not accurate, as it was necessary for the disease to be in an advanced stage to be detected by human olfaction. Based on this, electronic systems have been developed to diagnose diseases through the breath and to provide information about the state of the human body [1]. Human breath has a considerable amount of volatile organic compounds (VOCs) that are a product of metabolic activity. These VOCs can differ according to genetic or environmental factors such as age, weight, sex, lifestyle, or eating habits, and can influence the chemical composition of a person's breath, depending on the amount and concentrations of these compounds. Diseases can also cause an alteration of VOCs in the breath [2].

Currently, there has been an exponential development of technology in medical applications, in both the prevention and diagnosis or treatment of disorders. The growth in areas such as chemical sensors, microelectronic designs, material sciences, and artificial intelligence is contributing to the development of medical technology, and to improved health and increased life expectancy of the population. According to the World Health Organization (WHO), the average life expectancy in the world increased by five years in the period 2000–2016 [3]. The problem is that quality of life has not improved as much as life expectancy. The average increase in life years does not mean that this corresponds to a good state of health. People can suffer diseases and problems that cause a loss of quality of life without endangering their lives. On the other hand, health inequalities persist among different countries and people at different income levels [4].

One of the biggest challenges of modern medicine is to develop equipment that allows early diagnosis in a noninvasive way. This could avoid disease exacerbation, permitting control of the evolution of chronic patients in the medium and long term and maintaining their quality of life. It also would contribute to a reduction in hospital costs, as it could reduce the number of hospital/health center visits and the amount of medicine used. The principal limitation of the early detection of disease is that it is impossible to analyze healthy persons continuously, because this would be economically infeasible. In addition, the health system does not have the resources or the medical staff for this purpose. For this reason, the devices should be cheap and portable, so they can be available to more hospitals and health centers. On the other hand, this would open the possibility for patients to have devices at home to perform regular analysis.

These arguments and necessities have motivated the development of new technologies such as the electronic nose. The first electronic nose was developed in 1964 by Wilkens et al. [5], but it was not until 1982 when the electronic nose as a system using chemical sensors to classify odors was described by Persaud et al. [6]. In the 1980s, there was an evolution of this technology, which led to the creation of a technological sector that tried to imitate the olfactory systems of mammals [7]. Since then, the development of electronic noses has gone hand-in-hand with the technology being used in various sectors and different applications [8]. One is the medical sector, particularly the diagnosis and control of respiratory and digestive diseases. This technology could offer the possibility of diagnosing or evaluating disease states in a noninvasive and quick way with low-cost instruments.

2. The Olfactory Organ and Electronic Nose

Living organisms receive information about the surrounding environment through different sensory organs. Nowadays, the scientific community is focusing on the development and generation of systems and devices that can mimic the sensory organs: first, to replace the function of one of these organs in case of malfunction due to atrophy, pathology, or accident; and second, to be used in a wide variety of industrial applications in fields such as medicine, agriculture, food, environment, etc. Given that a deficiency in the ability to smell does not limit a person's normal life, except for the detection of gas leaks or fires, this technology becomes increasingly important in other applications [9,10].

The electronic nose is capable of detecting, discriminating, and identifying different types of chemicals present in the headspace of a sample as VOCs. The device's response to smell is produced by the interaction between volatile compounds and sensors, whereas this function is done in the biological nose by the olfactory epithelium, which works as a transductor, as it generates electrical signals from chemical stimuli. These signals are preprocessed in the olfactory bulb, then transmitted to the brain, where they are stored. Finally, the data are used to identify odors in the learning stage. Analogously, in an electronic nose, an analogical–digital conversion is produced to preprocess the signals, with a microcontroller employed for this purpose. In this case, the data are stored in the database of a pattern recognition machine to identify the aromas that are learned. Figure 1 shows a comparison between the electronic nose and biological nose [10,11].

Due to the wide variety of applications that have been generated recently in several sectors, sensors based on different detection principles have been developed to satisfy the needs that have arisen. Sensors undergo a physical or chemical change by interacting at the molecular level when exposed to a gas. This process is reversible and allows the sensors to be used again in other tests [8]. A wide variety of sensors have been developed: metal oxide semiconductor (MOS), conducting polymer (CP), chemocapacitor (CAP), electrochemical (EC), metal oxide semiconductor field effect (MOSFET), surface acoustic wave (SAW), quartz crystal microbalance (QCM), bulk acoustic wave (BAW), fluorescence (FL), optical fiber live cell (OF-LC), catalytic field effect (CFET), calorimetric or catalytic bead (CB), carbon black composite (CBC), micro-electromechanical system (MEMS), photoionization detector (PID), and amperometry gas sensor (AGS) [12].

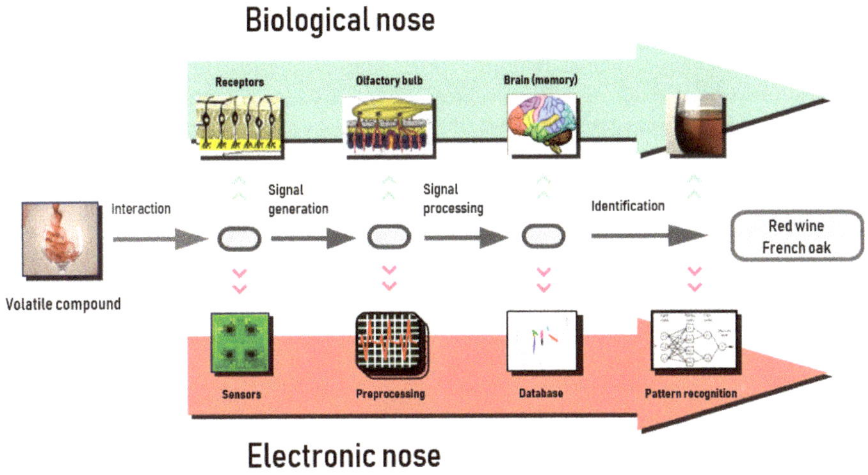

Figure 1. Schematics of electronic nose and biological olfactory system [13].

This offers a wide range of possibilities, so it is possible to use one type of sensor or another according to the application [8,12]. For example, sensors with good sensitivity are MOS, MOSFET, and AGS. SAW, FL, and OF-LC combine sensibility with specificity. However, the principal disadvantages of the gas sensor are slow recovery (MOS), drift in the response (MOS, CP, SAW), low noise immunity (PID), and lack of reproducibility between sensors of different sets (CP, MOSFET, QCM, SAW) and in the response of the same sensor in the long term. Because of this, it is normal to use a matrix of sensors of different types to avoid the disadvantages presented by each one separately and maximize their advantages [12,14].

3. Biomarkers and Diseases

In 1899, Thomson, who was interested in measuring the mass/charge ratio of an electron, created the first instrument similar to a mass spectrometer (MC), improving on the work previously done by Wien. In 1941, Martin and Synge published a paper describing the liquid–liquid chromatographic partition. However, it was not until 1952 when chromatography took on its gas–solid version (James and Martin). In later years, equipment was developed that used both techniques, which eliminated the disadvantages that each one presented separately [15,16]. One of the main problems encountered in the analysis of VOCs was the impossibility of capturing these compounds, as they are very volatile. This inconvenience was solved with the emergence of solid-phase microextraction (SPME) [17]. Since then, new techniques have been developed, such as gas chromatography–mass spectrometry (GC-MS), proton transfer reaction–mass spectrometry (PTR-MS), selected ion flow tube–mass spectrometry (SIFT-MS), field asymmetric ion mobility spectrometry (FAIMS), time-of-flight mass spectrometry (TOF-MS), ion mobility spectrometry (IMS), liquid chromatography–mass spectrometry (LC-MS), and high-performance liquid chromatography (HPLC).

With these techniques, it is possible to extract information about the respiratory or digestive system through the breath. However, the existence of VOCs in respiration does not necessarily imply that these volatile molecules are produced by the human body; rather, they can be produced in an exogenous process. For example, acetonitrile is commonly found in the breath of smokers, occurring exogenously. Exposure to a dangerous atmosphere, contaminants, or even certain medications can generate new compounds and also alter the concentration of other endogenous compounds [18].

Table 1 shows descriptions of different processes associated with oxidative stress and airway inflammation and the influence of these processes on the composition of VOCs in the breath. Endogenous compounds are produced by respiratory or digestive system cells, which metabolize the molecules present in the inspired air and generate others. When the cells function abnormally, they can alter the composition of VOCs in the breath. For example, oxidative stress is an imbalance of the normal state of the human body caused by the production of free radicals, which alters the composition of the compounds present in the breath as H2O2, CO, or 8-isoprostane [19,20]. Another example is the inflammation of airways, where the immune response produces an increase of various biomarkers in the breath [19,21].

Table 1. Processes involved in different pathologies.

Process	Description	Biomarker	References
Marker of oxidative stress	Inflammation process in lung cells; eosinophils, neutrophils, and macrophages produce reactive oxygen species	H_2O_2	[19,22,23]
	Increase of free radicals, which react to cell membrane phospholipid, generating 8-isoprostane	8-isoprostane	[19,20]
	Oxidation of cell membrane phospholipids produces a chain reaction, the targets of which are polyunsaturated fatty acids, resulting in the formation of unstable lipid hydroperoxides and secondary carbonyl compounds, such as aldehydic products	Malondialdehyde	[19]
	CO, a marker of oxidative stress, is produced by the stress protein hemoglobin oxygenase	CO	[21]
Inflammation of airways	Immune response against infection produces an inflammation process in cells, which generate more NO than in a healthy person	Alveolar NO	[19]
	Cytokines * and chemokines are involved in many aspects of the disease process in chronic obstructive pulmonary disease (COPD), including recruitment of neutrophils, macrophages, T cells, and B cells	Cytokines * and chemokines	[19]
	Leukotrienes are muscle constrictors, such as in lung muscle	Leukotriene B4 and prostaglandins	[19]
	CO is a marker of inflammation	CO	[21]

* Cytokines are agents responsible for cellular communication.

Microorganisms that may be present in the digestive or respiratory tract, such as viruses, bacteria, or fungi, can also directly or indirectly alter the composition and concentration of VOCs in the breath. These cannot modify the concentration by themselves; otherwise, these microorganisms can affect the normal functioning of the cells of the body, causing the same effect. Therefore, the alteration of volatile compounds can have an exogenous or endogenous cause, which can be produced by a malfunction of the cells or by external biological agents [18].

3.1. Endogenous Biomarkers

The following section describes the pathologies detected by different chemical techniques of analysis that are mentioned in the bibliography consulted. Asthma and chronic obstructive pulmonary disease (COPD) are chronic diseases that cause difficulty breathing due to inflammation of the airways, which become rigid [24,25]. Inflammation of the epithelial cells causes increased concentrations of NO, pentane, isoprene, and ethane in the expired air in asthma (Lärstad et al. [26]) and ethane in the expired air in COPD (Paredi et al. [27]) (Table 2). This occurs with other diseases described in the bibliography, such as acute respiratory distress syndrome (ARDS), cystic fibrosis (CF), and lung cancer. ARDS is a pathology that prevents enough oxygen from reaching the lungs and the bloodstream [28]. CF is a genetic and chronic disorder that affects organs such as the pancreas, liver, kidneys, and intestine, causes difficulty breathing, and generates dense mucus in the airways [29]. A study by Antuni et al. [30] used NO and CO as biomarkers for the discrimination of this disease. In the case of NO, the concentration of this biomarker decreases in patients with CF in comparison to healthy individuals. In the case of CO, the concentration increases.

Finally, in lung cancer, the process of cell division is disrupted by the creation of new cells in an uncontrolled way and when they are not necessary, as well as allowing old or damaged cells to survive [31]. Bajtarevic et al. [32] employed isoprene, acetone, methanol, and benzene as biomarkers of lung cancer. The concentrations of these compounds decreased in patients with this disease in comparison with healthy persons. Table 2 shows chemical analysis techniques and biomarkers used to detect different diseases.

Table 2. Concentrations of biomarkers used in the detection of different diseases. NS, not stated; ppb, parts per billion; ppmv, parts per million by volume.

Disease	Study	Biomarker	Concentration	References
Asthma	Lärstad (2007)	Ethane	NS	[26]
		NO	19 ± 2 ppb (healthy subject); 30 ± 6.1 ppb (asthma patient)	
		Pentane	NS	
		Isoprene	113 ppb	
	Olopade (1997)	Pentane	Acute asthma: 8.4 ± 2.9 nmol/L	[33]
		Pentane	Stable asthma: 3.6 ± 0.4 nmol/L	
	Paredi (2000)	Ethane	Ethane: asthma not treated with steroids: 2.06 ± 0.30 ppb; asthma treated with steroids: 0.79 ± 0.1 ppb); healthy volunteers: 0.88 ± 0.09 ppb NO: asthma not treated with steroids: 14.7 ± 1.7 ppb; asthma treated with steroids: 8.6 ± 0.5 ppb	[27]
	Dweik (2011)	NO	Low asthma patients: <25 ppb in adults; >20 ppb in children Intermediate asthma patients: 25–50 ppb in adults; 20–35 ppb in children High asthma patients: >50 ppb in adults; >35 ppb in children Persistently high asthma patients: >50 ppb in adults; 35 ppb in children	[34]
COPD	Paredi (2000)	Ethane	2.77 ± 0.25	[35]
Cystic fibrosis	Barker (2006)	Pentane	0.36 (0.24–0.48) ppb	[36]
		Dimethyl Sulfide	3.89 (2.24–5.54) ppb	
	Antuni (2000)	NO	Healthy volunteers: 7.3 (0.24) ppb; stable cystic fibrosis patients: 5.7 (00.29) ppb; unstable cystic fibrosis patients: 6.1 (0.72) ppb	[30]
		CO	Healthy volunteers: 2.0 (0.1) ppm; stable cystic fibrosis patients: 2.7 (0.22) ppm; unstable cystic fibrosis patients: 4.8 (0.3) ppb	
Lung cancer	Bajtarevic (2009)	Isoprene	Median concentration: healthy volunteers: 105.2 ppb; cancer patients: 81.5 ppb	[32]
		Acetone	Median concentration: healthy volunteers: 627.5 ppb; cancer patients: 458.7 ppb	
		Methanol	Median concentration: healthy volunteers: 142.0 ppb; cancer patients: 118.5 ppb	
		Benzene	Median concentration: healthy volunteers: 627.5 ppb; cancer patients: 458.7 ppb	

Table 2. *Cont.*

Disease	Study	Biomarker	Concentration		References
Diabetes mellitus	Das (2016)	Acetone	Type 1	0.044–2.744 ppm (healthy volunteers); 2.2–21 ppm (diabetes patients)	[37]
			Type 2	0.044–2.744 ppm (healthy volunteers); 1.76–9.4 ppm (diabetes patients)	
	Spanel (2011)	Acetone	Type 2	<800 ppb (healthy volunteers); >1760 ppb (diabetes patients)	[38]
Helicobacter pylori	Kearney (2002)	Dioxide carbon and ammonia.	NS		[39]
Hypolactasia	Metz (1975)	Hydrogen	Control: 0–3 ppmv; patients: 48–168 ppmv		[40]
Liver fibrosis	Alkhouri (2015)	Acetone	Lower fibrosis group: 117.8 ppb; advanced fibrosis group: 224.2 ppb		[41]
		Benzene	Lower fibrosis group: 1.9 ppb; advanced fibrosis group: 8 ppb		
		Carbon Disulfide	Lower fibrosis group: 1.6 ppb; advanced fibrosis group: 3.2 ppb		
		Isoprene	Lower fibrosis group: 13.5 ppb; advanced fibrosis group: 40.4 ppb		
		Pentane	Lower fibrosis group: 12.3 ppb; advanced fibrosis group: 19.5 ppb		
		Ethane	Lower fibrosis group: 63.0 ppb; advanced fibrosis group: 75.6 ppb		

As already mentioned, diseases of the digestive system, such as diabetes mellitus, liver fibrosis, and hypolactasia, can alter the concentration of VOCs in the same way as respiratory diseases. Diabetes is a disease in which the body does not produce insulin or does not use it correctly [42]. Liver fibrosis is a disorder that causes decreased blood supply through the liver and produces the accumulation of scar tissue [43]. Das et al. [37] and Kearney et al. [39] used acetone as the major biomarker to detect both pathologies. The concentration of this compound increased in patients with these disorders in comparison with healthy persons. Lastly, hypolactasia is a deficiency of lactase (the enzyme that metabolizes lactose) in the intestinal mucosa, which causes this molecule to not be metabolized [44]. For the detection of this disorder, Alkhouri et al. [41] used hydrogen as a biomarker.

3.2. Exogenous Biomarkers

In this section, diseases generated by external pathogens, such as pneumonia and pulmonary tuberculosis, which produce exogenous volatile organic compounds, are described. Pneumonia is a lung infection that can be caused by many pathogenic agents, such as bacteria, viruses, or fungi [45], while tuberculosis is a bacterial infection (Mycobacterium tuberculosis) that mainly attacks the lungs or other parts of the organism [46]. However, it should be noted that there are diseases (e.g., cystic fibrosis and COPD) that predispose the patient to infections such as pneumonia [29]. Therefore, in this pathology it is common to find endogenous and exogenous VOCs in the breath. In order to determine whether a person is ill, it is necessary to know the aromatic profile of a healthy person. The concentrations of existing major compounds in the breath of healthy persons can be seen in Table 3.

Table 3. Major volatile organic compounds present in the breath of healthy individuals [47].

Compound	Concentration
Water vapor	5–6.3%
Nitrogen	78.04%
Oxygen	16%
Carbon dioxide	4–5%
Hydrogen	5%
Argon	NS
CO	0–6 ppm
Ammonia	0.5–2 ppm
Acetone, methanol, ethanol	0.9%; <1 ppm
Hydrogen sulfide	0–1.3 ppm
NO	10–50 ppb
Carbonyl sulfide	0–10 ppb
Ethane	0–10 ppb
Pentane	0–10 ppb
Methane	2–10 ppm

4. Traditional Methods of Diagnosis

Traditionally, chemical analysis techniques have not been used to diagnose respiratory or digestive diseases, as these are very expensive. Breath analysis may mitigate some of the disadvantages of conventional diagnostic tests. Additionally, it could complement conventional methods as a screening tool. The main disadvantage of traditional diagnostic tests is that there is usually a long wait to conduct the test. In addition, some are invasive tests, which may require irradiation of the patient or a surgical procedure. In other cases, the test result is not obtained immediately, as in the case of cultures.

Both invasive and noninvasive diagnostic techniques are currently used. For example, spirometry and fractional exhaled nitric oxide (FeNO) are noninvasive techniques used to determine the patient's lung capacity, mainly used to diagnose diseases such as asthma and COPD. The sweat test and sputum cultures are also employed as noninvasive techniques for the diagnosis of cystic fibrosis (described by Gibson and Cooke [48] in 1959) and respiratory infections, respectively. On the other hand, diagnostic tests used to detect other diseases described in the bibliography are lung biopsy (lung cancer and lung sarcoidosis), chest x-ray (CXR) (ARDS and lung tuberculosis), endoscopy (lung cancer), and computed tomography (CT) (lung sarcoidosis and lung cancer).

5. Recent Developments in Electronic Noses for the Diagnosis of Respiratory Diseases

One of the major objectives in medicine today is to develop equipment and techniques to achieve early diagnosis of diseases. For this, it is necessary to develop devices that are portable, economical, and, above all, noninvasive, in order to reduce the risk to patients. In addition, this would contribute to decentralizing medical resources and facilitate access to these devices, reducing waiting lists. Until now, noninvasive tests that have been performed were based only on the chemical analysis of body fluids (feces, urine, and blood), but the possibility of analyzing the volatile compounds generated by the samples (as well as exhaled air) has not been explored.

Traditionally, chemical analysis techniques such as GC-MS and, to a lesser extent, the techniques described in Section 3 have been used to detect VOCs and the chemical composition of gaseous samples. With the evolution of gas sensors and the electronic nose in recent decades, they are beginning to be used as diagnostic methods. It is true that currently gas sensors are not as selective as chemical analysis equipment, but the electronic nose allows for much smaller and compact equipment, which offers a lot of possibilities.

This paper includes a summary of the specialized bibliography on the detection of respiratory and digestive diseases through the breath using an electronic nose (Table 4). It should be noted that the scientific research carried out in recent years is included here. Therefore, only the electronic noses used in such studies are described, with the intention that this paper may help to determine the current state of research in the field and be a starting point for future studies. In the bibliography, it is possible to find some authors who have used commercial electronic noses and others who have developed their own equipment adapted to specific applications. It should be noted that the diagnostic tests employed are used to corroborate the results of the studies, but they do not have to be standard tests. Most of the authors described in this paper used a commercial electronic nose, the Cyranose 320, which has a matrix of 32 sensors of carbon black polymer. Only a few used a self-developed electronic nose optimized for a specific application.

Asthma is one of the respiratory diseases that affects the largest number of people in the world, based on more papers being found on this disease. The population studied by Dragonieri et al. [49,50] was 40 adult subjects, and they employed the Cyranose 320. In data processing, they used principal component analysis (PCA) as an alternative diagnostic test to corroborate the results obtained by using spirometry and FeNO. The other authors studying asthma used similar population sizes, and also used the Cyranose 320. They used spirometry or FeNO as a diagnostic test. While not all authors used the same techniques in the processing phase, they employed PCA and other techniques such as receiver operating characteristic (ROC) curve and neural network algorithms (ANNs). On the other hand, Brinkman et al. [51] studied a smaller population of subjects and performed the measurements using three electronic noses, two commercial (Cyranose 320, carbon black polymer sensors) and Tor Vergata Electronic Nose (QCM sensors; developed by the Sensors Group at the University of Rome Tor Vergata, along with the Departments of Electronic Engineering and Chemical Science and Technology), and one home-developed incorporating metal oxide semiconductor sensors.

In the case of COPD, a variety of scientific articles were found that used the same two commercial electronic noses as in the previous case: Cyranose 320 and Tor Vergata Electronic Nose. Dymerski et al. [52] used an electronic nose formed by commercial sensors from Figaro. In this case, the samples were generated in the laboratory, unlike in the other studies. To distinguish between patients and healthy individuals, Paredi et al. [35] employed three biomarkers: ethanol, CO, and NO, the concentrations of which in the exhaled air of patients with COPD increased compared to the healthy people. The literature found for lung cancer varies in terms of the type of cancer described; each author focused on the detection of one, studying populations of between 50 and 100 adult patients, using the commercial electronic noses that were used by other authors.

In the detection of pulmonary tuberculosis, a wider variety of authors were found to use different electronic noses with different types of sensors: commercial ones such as Cyranose 320, Aeonose, and Bloodhound BH1114, and home-developed equipment, such as Zelota et al. [53] using QCM, Bruins et al. [54] using metal oxide semiconductor sensors, and Fend et al. [55] using conducting polymer sensors.

For the rest of the pathologies shown in the table, fewer bibliographical references were found. Saasa et al. [56] discriminated between samples of patients with diabetes mellitus by detecting acetone. On the other hand, Schnabel et al. [57] stand out, discriminating between patients with pneumonia and healthy people. They studied a remarkable sample of 125 adult patients using a commercial electronic nose called DiagNose, which uses metal oxide sensors.

Table 4. Biomedical applications developed using commercial and experimental electronic noses.

Application	Author	Population Characteristics	Sensor Technology	Number of Sensors	Data Processing Algorithm	Diagnosis	Other Techniques	References
Asthma	Dragonieri (2007)	40 adult subjects, nonsmokers, aged 18–75, without any other acute or chronic disease besides asthma (mixed group) • Group 1: 10 patients, 25.1 ± 5.9 years, intermittent-mild asthma • Group 2: 10 patients, 26.8 ± 6.4 years (control group) • Group 3: 10 patients, 49.5 ± 12.0 years, moderate-severe persistent asthma • Group 4: 10 patients, 57.3 ± 7.1 years (control group)	Polymer nanocomposite sensor	32	PCA	Spirometry, FeNO	GC-MS	[49]
	Montuschi (2010)	52 adult subjects, nonsmokers (mixed group) • Group 1: 27 asthma patients, 39 ± 3 years • Group 2: 24 patients, 33 ± 3 years (control group)	QCM gas sensors coated by molecular metalloporphyrin film	8	PCA and FNN	FeNO	GC-MS	[58]
	Santonico (2014)	58 subjects	Carbon black polymer (Cyranose C320)/QCM covered with metalloporphyrin film (Tor Vergata Electronic Nose)/metal oxide semiconductor	32/NS/NS	ROC	FeNO	FAIMS (Owlstone)	[59]
	Brinkman (2017)	28 subjects • 23 asthma patients, 25 (21–31) years • 5 healthy volunteers, control group	Carbon black polymer (Cyranose C320)/QCM covered with metalloporphyrin film (Tor Vergata Electronic Nose)/metal oxide semiconductor	32/NS/NS	PCA	FeNO and spirometry	FAIMS (Owlstone) and GC-MS	[51]
	Cavaleiro (2018)	60 subjects, aged 6–18 years (mixed group)	Polymer nanocomposite sensor (Cyranose 320)	32	Clustering	FeNO and spirometry	GC-MS	[60]

Table 4. *Cont.*

Application	Author	Population Characteristics	Sensor Technology	Number of Sensors	Data Processing Algorithm	Diagnosis	Other Techniques	References
Chronic obstructive pulmonary disease	Paredi (2000)	36 subjects (mixed group) • Group 1: 12 nonsteroid-treated patients, 60 ± 18 years • Group 2: 10 steroid-treated patients, 58 ± 2 years • Group 3: 14 healthy subjects, 33 ± 3 years (control group)	NS	NS	NS	NS	NS	[35]
	Capuano (2010)	20 subjects (mixed group) • Group 1: 12 COPD patients, ex-smokers, therapy not based on cortisone • Group 2: 8 healthy volunteers (control group)	QMC gas sensor with metalloporphyrin films	7	PLS-DA	NS	NS	[61]
	Hattesohl (2011)	33 subjects (mixed group) • Group 1: 10 COPD patients • Group 2: 23 patients (control group)	Polymer nanocomposite sensor (Cyranose 320)	32	LDA, MD, CVVs, canonical plot	Spirometry	GC-MS	[62]
	Dymerski (2013)	In vitro experiments	SAW/BAW sensors (TGS 880, TGS 825, TGS 826, TGS 822, TGS 2610, TGS 2602 by Figaro)	6	PCA	NS	NS	[52]
	Bofan (2013)	24 adult subjects, ex-smokers, 68 ± 1.7 years, with smoking history of 39.5 (24.2 –63.3) years without other acute or chronic disease besides COPD or nonatopic COPD and without inhaled or oral corticosteroids (mixed group)	Polymer nanocomposite and inorganic conductor sensor (carbon black) (Cyranose 320)	32	Pattern recognition algorithm	Spirometry and FeNO	GC-MS, NMR spectroscopy, and LC-MS	[63]
Acute respiratory distress syndrome	Bos (2014)	180 subjects (mixed group) • Group 1: 58 ARDS patients, 57 (54–78) years • Group 2: 11 pneumonia patients, 56 (49–62) years • Group 3: 19 cardiogenic pulmonary edema patients, 71 (63–79) years • Group 4: 92 healthy volunteers, 64 (50–75) years (control group)	Polymer nanocomposite sensor	32	ROC	CXR	GC-MS	[64]

Table 4. *Cont.*

Application	Author	Population Characteristics	Sensor Technology	Number of Sensors	Data Processing Algorithm	Diagnosis	Other Techniques	References
Pulmonary sarcoidosis	Dragonieri (2013)	31 subjects (mixed group) • Group 1: 11 patients, 48.4 ± 9.0 years, untreated pulmonary sarcoidosis • Group 2: 20 patients, 49.7 ± 7.9 years, treated pulmonary sarcoidosis • Group 3: 25 patients, 39.6 ± 14.1 years (control group)	Polymer nanocomposite sensor (Cyranose 320)	32	PCA, CDA ROC curves	CXR, CT, biopsy	GC-MS, TOF-MS, IMS	[65]
Cystic fibrosis	Paff (2013)	48 subjects (mixed group) • Group 1: 25 patients, 11.4 years • Positive bacterial cultures: 15/25 patients • Pulmonary exacerbation: 9/25 patients • Group 2: 23 patients, 9.3 years (control group)	Polymer nanocomposite sensor (Cyranose 320)	32	PCA, ROC curves, and CDA	Spirometry and sputum culture	GC-MS	[66]
Primary ciliary dyskinesia	Paff (2013)	48 subjects (mixed group) • Group 1: 25 patients, 10.7 years • Positive bacterial cultures: 4/25 patients • Pulmonary exacerbation: 8/25 patients • Group 2: 23 patients, 9.3 years (control group)	Polymer nanocomposite sensor (Cyranose 320)	32	PCA, ROC curves, and CDA	Spirometry and sputum culture	GC-MS	[66]

Table 4. *Cont.*

Application	Author	Population Characteristics	Sensor Technology	Number of Sensors	Data Processing Algorithm	Diagnosis	Other Techniques	References
	Di Natale (2003)	50 subjects (mixed group) • Group 1: 42 patients with various forms of cancer not showing any other disease • Group 2: 8 patients without respiratory disease not taking any medication	QCM gas sensors coated with metalloporphyrin molecular film	8	PLS-DA	NS	GC–MS	[67]
	Machado (2005)	59 subjects (mixed group) • Group 1: 14 patients, 64 ± 3 years, with untreated bronchogenic carcinoma • 13 patients with non-small-cell cancer • 1 patient with small-cell cancer • Group 2: 45 patients, 38 ± 2 years (control group)	Polymer nanocomposite sensor (Cyranose 320)	32	PCA, SVM, and CDA	CXR and biopsy	GC–MS	[68]
Lung cancer	Dragonieri (2009)	30 subjects (mixed group) • Group 1: 10 NSCLC patients, 66.4 ± 9.0 years: 2 current smokers, 7 ex-smokers, 1 never smoked • Group 2: 10 COPD patients, 61.4 ± 5.5 years: 6 current smokers, 4 ex-smokers • Group 3: 10 healthy volunteers, 58.3 ± 8.1 years, never smoked (control group)	Polymer nanocomposite sensor (Cyranose 320)	32	CDA, CVV, PCA	CT	GC–MS	[69]
	Dragonieri (2012)	39 subjects (mixed group) • Group 1: 13 patients, 60.9 ± 12.2 years, with confirmed diagnosis of MPM • Group 2: 13 subjects, 67.2 ± 9.8 years, long-term professional exposure to asbestos • Group 3: 13 subjects, 52.2 ± 16 years, no asbestos exposure (control group)	Polymer nanocomposite sensor (Cyranose 320)	32	CVA, PCA, and ROC	NS	GC–MS	[70]

Table 4. *Cont.*

Application	Author	Population Characteristics	Sensor Technology	Number of Sensors	Data Processing Algorithm	Diagnosis	Other Techniques	References
Lung cancer	D'Amico (2010)	98 adult subjects, 50–70 years (mixed group) • Group 1: 56 patients with primary lung cancer, ex-smokers, not under oncological therapy, at least 6 months from last intervention • Group 2: 36 patients with normal lung function, negative history of chest symptoms, nonsmokers, no history of respiratory disease (control group)	QCM gas sensors	8	PLS-DA	Endoscopy	GC-MS	[23,71]
Pneumonia	Hockstein (2005)	25 subjects (mixed group) • Group 1: 13 patients with diagnosed pneumonia • Group 2: 12 patients without pneumonia	Polymer nanocomposite sensor (Cyranose 320)	32	SMV and PCA	CT	GC–MS	[72]
	Chiu (2015)	In vitro experiment	Polymer–carbon composite with polymers on chemical sensor array	8	CRBM	CXR, blood draw, and sputum culture	GC–MS	[73]
	Schnabel (2015)	125 subjects (mixed group) • Group 1: 33 pneumonia patients, 62 (20–82) years, subject to BAL • Group 2: 39 pneumonia patients, 57 (23–82) years • Group 3: 53 patients, 60 (34–85) (control group)	MOS sensors (DiagNose)	NS	PCA and ROC curves	CT	GC–MS	[57]
Pulmonary tuberculosis	Pavlou (2004)	In vitro experiment	Gas-sensor array (Bloodhound BH114)	14	PCA, optimization of BP-FNN, multivariate techniques,	CXR	NS	[74]

Table 4. *Cont.*

Application	Author	Population Characteristics	Sensor Technology	Number of Sensors	Data Processing Algorithm	Diagnosis	Other Techniques	References
Pulmonary tuberculosis	Fend (2006)	330 patients (mixed group) • 188 pulmonary tuberculosis patients: 53.7% HIV patients, 31.4% smokers • 142 nonpulmonary tuberculosis patients: 29.6% HIV patients, 9.2% smokers	CP sensor	14	PCA, DFA, and ANN	CXR and sputum culture	GC-MS	[55]
	Bruins (2013)	30 patients (mixed group). • Group 1: 15 pulmonary tuberculosis patients, 32 (21–58) years • Group 2: 15 healthy volunteers, 30 (18–58) years (control group)	MOS sensor: AS-MLC, AS-MLN, AS-MLK, AS-MLV	12 (4 types of sensors in triplicate)	ANN	CXR and microbiological culture	GC-MS	[54]
	Coronel (2017)	110 subjects (mixed group) • Group 1: 47 pulmonary tuberculosis patients, 34.6 years • Group 2: 14 COPD or asthma patients, 54.5 years • Group 3: 49 patients (control group)	MOS sensors (Aeonose)	NS	ROC curve	CXR	GC-MS	[75]
	Zelota (2017)	71 subjects (mixed group) • Group 1: 31 pulmonary tuberculosis and HIV patients, 28.7 ± 7.2 years • Group 2: 20 pulmonary tuberculosis patients without HIV, 39 ± 9.3 years • Group: 20 healthy volunteers, 33 ± 11 years (control group)	QCM gas sensors coated by metalloporphyrin molecular film	8	PCA	CXR	GC-MS	[53]
	Mohamed (2017)	500 patients (mixed group) • Group 1: 260 pulmonary tuberculosis patients, 41.72 ± 16.03 years • Group 2: 204 healthy volunteers, 43.38 ± 12.42 years (control group)	MOS sensor	10	PCA and ANN	Physical examination and routine laboratory analyses, including CXR	GC-MS	[76]
Diabetes mellitus	Saasa (2018)	NS	QCL, LAP, and chemoresistive sensors	NS	NS	NS	GC-MS, LC-MS, HPLC, PTR-MS, and SIFT-MS	[56]

6. Future and Challenges

Commercial electronic noses, used by most researchers, are designed for general use and are not optimized to detect the biomarkers of the diseases of interest. For medical applications, it is necessary to design specific sensors and optimized electronic noses, since it is an extremely complex sector, and it is a critical factor for the diagnosis and control of pathologies to be effective. In the medical sector, it is necessary to develop equipment that allows early diagnosis of diseases, enables simple and effective control, and is noninvasive.

On the one hand, this would allow health centers and small hospitals to have access to this type of equipment for the diagnosis and control of diseases (due to its lower cost) and thus minimize waiting lists. On the other hand, this would provide an opportunity for patients to have devices at home. In addition, in conjunction with the evolution of mobile technologies and Internet communication, it would allow patients to diagnose and control diseases from home by themselves or with remote medical supervision. It could even permit these types of sensors to be connected to phones either by incorporating them or through a gadget. The tendency of medicine is to advance along this line so that analyses and explorations can be performed at a distance. It is hospitals and health insurers themselves that could provide these types of devices to patients. In this way, it would be possible to achieve a unified protocol of measurement from the breath that is simple for patients to perform.

It is a promising technology for the diagnosis of respiratory and digestive diseases in a noninvasive way, since it allows the development of portable and cost-effective equipment. This is a key factor for this technology to succeed. However, current sensors still present complications that limit their operation. Therefore, one of the challenges from a technological point of view is to improve the selectivity of these sensors and their sensitivity to lower concentrations. To this end, it is important to use and develop new nanomaterials that can significantly improve selectivity and reduce the size and consumption, essential issues for miniaturization. It is therefore perhaps more important to focus efforts on developing technology that would enable these devices to be used in this application, and then study which is the ideal methodology to take samples in patients that is easy for a person without medical education to do.

7. Conclusions

Electronic nose technology has evolved remarkably in the last decade. Many researchers have focused their efforts on developing this type of equipment in a multitude of sectors, such as food, medicine, environment, and detection of hazardous materials.

This study presents a bibliographic review of research done on the use of the electronic nose for the detection of respiratory and digestive diseases through the breath. This technology bases its design and operation on the human sensory organ. The e-nose technology permits the study of diseases from a new approach, since until now analytical tests have focused exclusively on blood, urine, or feces samples, which allowed additional information to be obtained and studied. Although this technology is still in development and currently has some limitations, it has several advantages over other diagnostic methods. The ability to take measurements through the breath makes it possible to conduct analyses noninvasively, eliminating risks to the patient.

In addition, it would allow the production of portable equipment at a lower price than other equipment available in the market, giving health centers and hospitals greater access to this equipment. With the evolution in recent years and current development (improvement and miniaturization of gas sensors), the electronic nose is presented as a promising technology that should contribute to improving the quality of life of patients with chronic pathologies as well as early diagnosis of different diseases, contributing to the reduction of direct and indirect costs of the health system.

The purpose of this paper is to demonstrate the viability of this technology through a large number of studies done in this regard. From now on, it will be necessary to improve and solve the problems presented by the e-nose, fine-tuning this technology for the different applications in which it is intended to be used. This will require a great effort on the part of the research community, but once these problems have been solved, this technology should result in a great advance in the control and diagnosis of diseases, and satisfy many other needs that currently exist in the medical sector.

Funding: This research was funded by Department of Education and Research of the Madrid Autonomous Community (Spain) grant number IND2017/TIC7714.

Conflicts of Interest: The authors declare no conflict of interest.

Abbreviations

ANN	Artificial neural network
AS-MLC	Metal oxide semiconductor sensor for detection of carbon monoxide, manufactured by Applied Sensor Technologies
AS-MLK	Metal oxide semiconductor sensor, manufactured by Applied Sensor Technologies
AS-MLN	Metal oxide semiconductor sensor for detection of nitrogen monoxide sensor, manufactured by Applied Sensor Technologies
AS-MLV	Metal oxide semiconductor sensor for detection and reduction of gases such as VOCs and CO, manufactured by Applied Sensor Technologies
BAL	Bronchoalveolar lavage
BP	Backpropagation
CDA	Canonical discriminant analysis
CO	Carbon monoxide
CRBM	Convolutional restricted Boltzmann machine (a type of probabilistic neural network)
CVA	Cross-validated accuracy
CVV	Cross-validation value
DFA	Discriminant function analysis
FNN	Feedforward neural network
GC	Gas chromatography
H_2O_2	Hydrogen peroxide
LAP	Light-addressable potentiometric
LTB4	Leukotriene B4
MD	Mahalanobis distance
MDA	Malondialdehyde
MS	Mass spectrometry
NMR	Nuclear magnetic resonance
NO	Nitric oxide
PCA	Principal component analysis
PLS-DA	Partial least squares–discriminant analysis
QCL	Quantum cascade laser
ROC	Receiver operating characteristic
SVM	Support vector machine

References

1. Professional Practice Committee: Standards of Medical Care in Diabetes-2018. *Diabetes Care* **2018**, *41*, S3. [CrossRef] [PubMed]
2. Breath Biopsy: The Complete Guide. Volatile Ogranic Compounds (VOCs) as Biomarkers for Disesases. Available online: https://www.owlstonemedical.com/breath-biopsy-complete-guide/ (accessed on 20 September 2018).
3. World Health Organization (WHO). Life Expectancy Increased by 5 Years Since 2000, but Health Inequalities Persist. Available online: http://www.who.int/news-room/19-05-2016-life-expectancy-increased-by-5-years-since-2000-but-health-inequalities-persist (accessed on 21 October 2018).
4. World Health Organization Word report on ageing and health. *Anim. Genet.* **2015**, *39*, 561–563.
5. Wilkens, W.F.; Hartman, J.D. An Electronic Analog the Olfactory Processes. *J. Food Sci.* **1964**, *29*, 372–378. [CrossRef]

6. Persaud, K.; Dodd, G. Analysis of discrimination mechanisms in the mammalian olfactory system using a model nose. *Nature* **1982**, *299*, 352–355. [CrossRef] [PubMed]

7. Gardner, J.W.; Bartlett, P.N. A brief history of electronic noses. *Sens. Actuators* **1994**, *18–19*, 210–211. [CrossRef]

8. Wilson, A.D. Diverse applications of electronic-nose technologies in agriculture and forestry. *Sensors* **2013**, *13*, 2295–2348. [CrossRef] [PubMed]

9. Ghasemi-Varnamkhasti, M.; Apetrei, C.; Lozano, J.; Anyogu, A. Potential use of electronic noses, electronic tongues and biosensors as multisensor systems for spoilage examination in foods. *Trends Food Sci. Technol.* **2018**, *80*, 71–92. [CrossRef]

10. Zhang, X.; Cheng, J.; Wu, L.; Mei, Y.; Jaffrezic-Renault, N.; Guo, Z. An overview of an artificial nose system. *Talanta* **2018**, *184*, 93–102. [CrossRef] [PubMed]

11. Santos, J.P.; Lozano, J.; Aleixandre, M. *Brewing Technology Chapter 9: Electronic Noses Applications in Beer Technology*; Kanauchi, M., Ed.; InTechOpen: Lodon, UK, 2017. [CrossRef]

12. Pearce, T.C.; Schiffman, S.S.; Nagle, H.T.; Gardner, J.W. *Handbook of Machine Olfaction*; John Wiley & Sons: Hoboken, NJ, USA, 2014; ISBN 047186109X.

13. Lozano, M.; Horrillo, M.; Aleixandre, M.C.; Santos, J.P.; Fernández, M.; Sayago, I.; Gutiérrez, J.L.; Fontecha, J. Olfative Sensor Systems for the Wine-Producing Industry. *Food* **2007**, *1*, 23–29.

14. Correa Hernando, E.C.; Barreiro Elorza, P.; Ruiz-Altisent, M.; Chamorro Valencia, M.C. Nariz electrónica ¿herramienta para la calidad en la industria agroalimentaria? «II Congreso Nacional de Calidad Alimentaria». Available online: http://www.agro-alimentarias.coop/ficheros/doc/01297 (accessed on 25 October 2018).

15. Adlard, E.R.; Poole, C.F. *Gas Chromatography: Historical Development*; Elsevier Inc.: Amsterdam, The Netherlands, 2018; ISBN 9780124095472.

16. Kolomnikov, I.G.; Efremov, A.M.; Tikhomirova, T.I.; Sorokina, N.M.; Zolotov, Y.A. Early stages in the history of gas chromatography. *J. Chromatogry A* **2018**, *1537*, 109–117. [CrossRef] [PubMed]

17. Vas, G.; Vékey, K. Solid-phase microextraction: A powerful sample preparation tool prior to mass spectrometric analysis. *J. Mass Spectrom.* **2004**, *39*, 233–254. [CrossRef] [PubMed]

18. Amann, A.; Costello, B.D.L.; Miekisch, W.; Schubert, J.; Buszewski, B.; Pleil, J.; Ratcliffe, N.; Risby, T. The human volatilome: Volatile organic compounds (VOCs) in exhaled breath, skin emanations, urine, feces and saliva. *J. Breath Res.* **2014**, *8*. [CrossRef] [PubMed]

19. Kostikas, K.; Bakakos, P.; Loukides, S. *General Methods in Biomarker Research and their Applications*; Preedy, V.R., Patel, V.B., Eds.; Springer: Dordrecht, The Ntherlands, 2014; pp. 1–25. [CrossRef]

20. Céspedes Miranda, E.; Castillo Herrera, J. La peroxidación lipídica en el diagnostico del estrés oxidativo del paciente hipertenso. ¿Realidad o mito? *Rev. Cubana Investig. Bioméd.* **2008**, *27*. Available online: http://scielo.sld.cu/scielo.php?script=sci_arttext&pid=S0864-03002008000200003 (accessed on 25 October 2018).

21. Shorter, J.H.; Nelson, D.D.; McManus, J.B.; Zahniser, M.S.; Sama, S.R.; Milton, D.K. Clinical study of multiple breath biomarkers of asthma and COPD (NO, CO_2, CO and N_2O) by infrared laser spectroscopy. *J. Breath Res.* **2011**, *5*, 037108. [CrossRef] [PubMed]

22. Phillips, M.; Cataneo, R.N.; Cummin, A.R.C.; Gagliardi, A.J.; Gleeson, K.; Greenberg, J.; Maxfield, R.A.; Rom, W.N. Detection of Lung Cancer With Volatile Markers in the Breath. *Chest* **2003**, *123*, 2115–2123. [CrossRef] [PubMed]

23. D'Amico, A.; Pennazza, G.; Santonico, M.; Martinelli, E.; Roscioni, C.; Galluccio, G.; Paolesse, R.; Di Natale, C. An investigation on electronic nose diagnosis of lung cancer. *Lung Cancer* **2010**, *68*, 170–176. [CrossRef] [PubMed]

24. Asthma: MedlinePlus Medical Encyclopedia. Available online: https://medlineplus.gov/ency/article/000141.htm (accessed on 20 October 2018).

25. COPD: MedlinePlus Medical Encyclopedia. Available online: https://medlineplus.gov/copd.html (accessed on 20 October 2018).

26. Lärstad, M.A.E.; Torén, K.; Bake, B.; Olin, A.C. Determination of ethane, pentane and isoprene in exhaled air—Effects of breath-holding, flow rate and purified air. *Acta Physiol.* **2007**, *189*, 87–98. [CrossRef] [PubMed]

27. Paredi, P.; Kharitonov, S.A.; Barnes, P.J. Elevation of Exhaled Ethane Concentration in Asthma. *Am. J. Respir. Crit. Care Med.* **2000**, *162*, 1450–1454. [CrossRef] [PubMed]

28. Acute Respiratory Distress Syndrome: MedlinePlus Medical Encyclopedia. Available online: https://medlineplus.gov/ency/article/000103.htm (accessed on 24 October 2018).

29. Cystic Fibrosis: MedlinePlus Medical Encyclopedia. Available online: https://medlineplus.gov/cysticfibrosis.html (accessed on 21 October 2018).
30. Antuni, J.D.; Kharitonov, S.A.; Hughes, D.; Hodson, M.E.; Barnes, P.J. Increase in exhaled carbon monoxide during exacerbations of cystic fibrosis. *Thorax* **2000**, *55*, 138–142. [CrossRef] [PubMed]
31. Lung Cancer: MedlinePlus Medical Encyclopedia. Available online: https://medlineplus.gov/lungcancer.html (accessed on 21 October 2018).
32. Bajtarevic, A.; Ager, C.; Pienz, M.; Klieber, M.; Schwarz, K.; Ligor, M.; Ligor, T.; Filipiak, W.; Denz, H.; Fiegl, M.; et al. Noninvasive detection of lung cancer by analysis of exhaled breath. *BMC Cancer* **2009**, *9*, 348. [CrossRef] [PubMed]
33. Olopade, C.O.; Zakkar, M.; Swedler, W.I.; Rubinstein, I. Exhaled pentane levels in acute asthma. *Chest* **1997**, *111*, 862–865. [CrossRef] [PubMed]
34. Dweik, R.A.; Boggs, P.B.; Erzurum, S.C.; Irvin, C.G.; Leigh, M.W.; Lundberg, J.O.; Olin, A.-C.; Plummer, A.L.; Taylor, D.R.; et al. An official ATS clinical practice guideline: interpretation of exhaled nitric oxide levels (FENO) for clinical applications. *Am. J. Respir. Crit. Care Med.* **2011**, *184*, 602–615. [CrossRef] [PubMed]
35. Paredi, P.; Kharitonov, S.A.; Leak, D.; Ward, S.; Cramer, D.; Barnes, P.J. Exhaled Ethane, a Marker of Lipid Peroxidation, Is Elevated in Chronic Obstructive Pulmonary Disease. *Am. J. Respir. Crit. Care Med.* **2000**, *162*, 369–373. [CrossRef] [PubMed]
36. Barker, M.; Hengst, M.; Schmid, J.; Buers, H.-J.; Mittermaier, B.; Klemp, D.; Koppmann, R. Volatile organic compounds in the exhaled breath of young patients with cystic fibrosis. *Eur. Respir. J.* **2006**, *27*, 929–936. [CrossRef] [PubMed]
37. Das, S.; Pal, S.; Mitra, M. Significance of Exhaled Breath Test in Clinical Diagnosis: A Special Focus on the Detection of Diabetes Mellitus. *J. Med. Biol. Eng.* **2016**, *36*, 605–624. [CrossRef] [PubMed]
38. Španěl, P.; Dryahina, K.; Rejšková, A.; Chippendale, T.W.E.; Smith, D. Breath acetone concentration; Biological variability and the influence of diet. *Physiol. Meas.* **2011**, *32*. [CrossRef] [PubMed]
39. Kearney, D.J.; Hubbard, T.; Putnam, D. Breath ammonia measurement in Helicobacter pylori infection. *Dig. Dis. Sci.* **2002**, *47*, 2523–2530. [CrossRef] [PubMed]
40. Metz, G.; Jenkins, D.J.; Peters, T.J.; Newman, A.; Blendis, L.M. Breath hydrogen as a diagnostic method for hypolactasia. *Lancet* **1975**, *1*, 1155–1157. [CrossRef]
41. Alkhouri, N.; Singh, T.; Alsabbagh, E.; Guirguis, J.; Chami, T.; Hanouneh, I.; Grove, D.; Lopez, R.; Dweik, R. Isoprene in the Exhaled Breath is a Novel Biomarker for Advanced Fibrosis in Patients with Chronic Liver Disease: A Pilot Study. *Clin. Transl. Gastroenterol.* **2015**, *6*, e112. [CrossRef] [PubMed]
42. Diabetes: MedlinePlus Medical Encyclopedia. Available online: https://medlineplus.gov/diabetes.html (accessed on 30 October 2018).
43. Bataller, R.; Brenner, D.A. Liver fibrosis. *J. Clin. Investig.* **2005**, *115*, 209–218. [CrossRef] [PubMed]
44. Lember, M. Hypolactasia: A common enzyme deficiency leading to lactose malabsorption and intolerance. *Pol. Arch. Med. Wewn.* **2012**, *122* (Suppl. 1), 60–64.
45. Pneumonia: MedlinePlus Medical Encyclopedia. Available online: https://medlineplus.gov/pneumonia.html (accessed on 31 October 2018).
46. Nahid, P.; Dorman, S.E.; Alipanah, N.; Barry, P.M.; Brozek, J.L.; Cattamanchi, A.; Chaisson, L.H.; Chaisson, R.E.; Daley, C.L.; Grzemska, M.; et al. Official American Thoracic Society/Centers for Disease Control and Prevention/Infectious Diseases Society of America Clinical Practice Guidelines: Treatment of Drug-Susceptible Tuberculosis. *Clin. Infect. Dis.* **2016**, *63*, e147–e195. [CrossRef] [PubMed]
47. Mathew, T.L.; Pownraj, P.; Abdulla, S.; Pullithadathil, B. Technologies for Clinical Diagnosis Using Expired Human Breath Analysis. *Diagnostics* **2015**, *5*, 27–60. [CrossRef] [PubMed]
48. Gibson, L.E.; Cooke, R.E. A test for concentration of electrolytes in sweat in cystic fibrosis of the pancreas utilizing pilocarpine by iontophoresis. *Pediatrics* **1959**, *23*, 545–549. [PubMed]
49. Dragonieri, S.; Schot, R.; Mertens, B.J.A.; Le Cessie, S.; Gauw, S.A.; Spanevello, A.; Resta, O.; Willard, N.P.; Vink, T.J.; Rabe, K.F.; et al. An electronic nose in the discrimination of patients with asthma and controls. *J. Allergy Clin. Immunol.* **2007**, *120*, 856–862. [CrossRef] [PubMed]
50. Dragonieri, S.; Quaranta, V.N.; Carratu, P.; Ranieri, T.; Resta, O. Exhaled breath profiling by electronic nose enabled discrimination of allergic rhinitis and extrinsic asthma. *Biomarkers* **2018**, 1–24. [CrossRef] [PubMed]

51. Brinkman, P.; van de Pol, M.A.; Gerritsen, M.G.; Bos, L.D.; Dekker, T.; Smids, B.S.; Sinha, A.; Majoor, C.J.; Sneeboer, M.M.; Knobel, H.H.; et al. Exhaled breath profiles in the monitoring of loss of control and clinical recovery in asthma. *Clin. Exp. Allergy* **2017**, *12*, 3218–3221. [CrossRef] [PubMed]

52. Dymerski, T.; Gebicki, J.; Wiśniewska, P.; Śliwińska, M.; Wardencki, W.; Namieśnik, J. Application of the electronic nose technique to differentiation between model mixtures with COPD markers. *Sensors* **2013**, *13*, 5008–5027. [CrossRef] [PubMed]

53. Zetola, N.M.; Modongo, C.; Matsiri, O.; Tamuhla, T.; Mbongwe, B.; Matlhagela, K.; Sepako, E.; Catini, A.; Sirugo, G.; Martinelli, E.; et al. Diagnosis of pulmonary tuberculosis and assessment of treatment response through analyses of volatile compound patterns in exhaled breath samples. *J. Infect.* **2017**, *74*, 367–376. [CrossRef] [PubMed]

54. Bruins, M.; Rahim, Z.; Bos, A.; Van De Sande, W.W.J.; Endtz, H.P.; Van Belkum, A. Diagnosis of active tuberculosis by e-nose analysis of exhaled air. *Tuberculosis* **2013**, *93*, 232–238. [CrossRef] [PubMed]

55. Fend, R.; Kolk, A.H.J.; Bessant, C.; Buijtels, P.; Klatser, P.R.; Woodman, A.C. Prospects for clinical application of electronic-nose technology to early detection of Mycobacterium tuberculosis in culture and sputum. *J. Clin. Microbiol.* **2006**, *44*, 2039–2045. [CrossRef] [PubMed]

56. Saasa, V.; Malwela, T.; Beukes, M.; Mokgotho, M.; Liu, C.-P.; Mwakikunga, B. Sensing Technologies for Detection of Acetone in Human Breath for Diabetes Diagnosis and Monitoring. *Diagnostics* **2018**, *8*. [CrossRef] [PubMed]

57. Schnabel, R.M.; Boumans, M.L.L.; Smolinska, A.; Stobberingh, E.E.; Kaufmann, R.; Roekaerts, P.M.H.J.; Bergmans, D.C.J.J. Electronic nose analysis of exhaled breath to diagnose ventilator-associated pneumonia. *Respir. Med.* **2015**, *109*, 1454–1459. [CrossRef] [PubMed]

58. Montuschi, P.; Santonico, M.; Mondino, C.; Pennazza, G.; Maritini, G.; Martinelli, E.; Capuano, R.; Ciabattoni, G.; Paolesse, R.; Di Natale, C.; et al. Diagnostic performance of an electronic nose, fractional exhaled nitric oxide, and lung function testing in asthma. *Chest* **2010**, *137*, 790–796. [CrossRef] [PubMed]

59. Santonico, M.; Zompanti, A.; Vernile, C.; Pennazza, G.; Brinkman, P.; Wagener, A.H.; Sterk, P.J.; D'Amico, A.; Montuschi, P. An investigation on e-nose platform relevance to respiratory diseases. In Proceedings of the IEEE Sensors, Valencia, Spain, 2–5 November 2014; pp. 688–690. [CrossRef]

60. Cavaleiro Rufo, J.; Paciencia, I.; Mendes, F.C.; Farraia, M.; Rodolfo, A.; Silva, D.; de Oliveira Fernandes, E.; Delgado, L.; Moreira, A. Exhaled breath condensate volatilome allows sensitive diagnosis of persistent asthma. *Allergy* **2018**. [CrossRef] [PubMed]

61. Capuano, R.; Santonico, M.; Martinelli, E.; Pennazza, G.; Paolesse, R.; Bergamini, A.; Cazzola, M.; Ciaprini, C.; Di Natale, C.; D'Amico, A. COPD diagnosis by a gas sensor array. *Procedia Eng.* **2010**, *5*, 484–487. [CrossRef]

62. Hattesohl, A.D.M.; Jörres, R.A.; Dressel, H.; Schmid, S.; Vogelmeier, C.; Greulich, T.; Noeske, S.; Bals, R.; Koczulla, A.R. Discrimination between COPD patients with and without alpha 1-antitrypsin deficiency using an electronic nose. *Respirology* **2011**, *16*, 1258–1264. [CrossRef] [PubMed]

63. Bofan, M.; Mores, N.; Baron, M.; Dabrowska, M.; Valente, S.; Schmid, M.; Trové, A.; Conforto, S.; Zini, G.; Cattani, P.; et al. Within-day and between-day repeatability of measurements with an electronic nose in patients with COPD. *J. Breath Res.* **2013**, *7*, 017103. [CrossRef] [PubMed]

64. Bos, L.D.J.; Schultz, M.J.; Sterk, P.J. Exhaled breath profiling for diagnosing acute respiratory distress syndrome. *BMC Pulm. Med.* **2014**, *14*, 72. [CrossRef] [PubMed]

65. Dragonieri, S.; Brinkman, P.; Mouw, E.; Zwinderman, A.H.; Carratú, P.; Resta, O.; Sterk, P.J.; Jonkers, R.E. An electronic nose discriminates exhaled breath of patients with untreated pulmonary sarcoidosis from controls. *Respir. Med.* **2013**, *107*, 1073–1078. [CrossRef] [PubMed]

66. Paff, T.; van der Schee, M.P.; Daniels, J.M.A.; Pals, G.; Postmus, P.E.; Sterk, P.J.; Haarman, E.G. Exhaled molecular profiles in the assessment of cystic fibrosis and primary ciliary dyskinesia. *J. Cyst. Fibros.* **2013**, *12*, 454–460. [CrossRef] [PubMed]

67. Di Natale, C.; Macagnano, A.; Martinelli, E.; Paolesse, R.; D'Arcangelo, G.; Roscioni, C.; Finazzi-Agrò, A.; D'Amico, A. Lung cancer identification by the analysis of breath by means of an array of non-selective gas sensors. *Biosens. Bioelectron.* **2003**, *18*, 1209–1218. [CrossRef]

68. Machado, R.F.; Laskowski, D.; Deffenderfer, O.; Burch, T.; Zheng, S.; Mazzone, P.J.; Mekhail, T.; Jennings, C.; Stoller, J.K.; Pyle, J.; et al. Detection of Lung Cancer by Sensor Array Analyses of Exhaled Breath. *Am. J. Respir. Crit. Care Med.* **2005**, *171*, 1286–1291. [CrossRef] [PubMed]

69. Dragonieri, S.; Annema, J.T.; Schot, R.; van der Schee, M.P.C.; Spanevello, A.; Carratú, P.; Resta, O.; Rabe, K.F.; Sterk, P.J. An electronic nose in the discrimination of patients with non-small cell lung cancer and COPD. *Lung Cancer* **2009**, *64*, 166–170. [CrossRef] [PubMed]

70. Dragonieri, S.; Van Der Schee, M.P.; Massaro, T.; Schiavulli, N.; Brinkman, P.; Pinca, A.; Carratú, P.; Spanevello, A.; Resta, O.; Musti, M.; et al. An electronic nose distinguishes exhaled breath of patients with Malignant Pleural Mesothelioma from controls. *Lung Cancer* **2012**, *75*, 326–331. [CrossRef] [PubMed]

71. American Cancer Society Estadísticas importantes sobre el cáncer de pulmón. Available online: https://www.cancer.org/es/cancer/cancer-de-pulmon-no-microcitico/acerca/estadisticas-clave.html (accessed on 4 September 2018).

72. Hockstein, N.G.; Thaler, E.R.; Lin, Y.; Lee, D.D.; Hanson, C.W. Correlation of pneumonia score with electronic nose signature: A prospective study. *Ann. Otol. Rhinol. Laryngol.* **2005**, *114*, 504–508. [CrossRef] [PubMed]

73. Chiu, S.; Member, S.; Wang, J.; Chang, K.; Chang, T.; Wang, C.; Chang, C.; Tang, C.; Chen, C.; Shih, C.; et al. A Fully Integrated Nose-on-a-Chip for Rapid Diagnosis of Ventilator- Associated Pneumonia. *IEEE Trans. Biomed. Circuits Syst.* **2015**, *8*, 2015. [CrossRef]

74. Pavlou, A.K.; Magan, N.; Jones, J.M.; Brown, J.; Klatser, P.; Turner, A.P.F. Detection of Mycobacterium tuberculosis (TB) in vitro and in situ using an electronic nose in combination with a neural network system. *Biosens. Bioelectron.* **2004**, *20*, 538–544. [CrossRef] [PubMed]

75. Coronel Teixeira, R.; Rodríguez, M.; Jiménez de Romero, N.; Bruins, M.; Gómez, R.; Yntema, J.B.; Chaparro Abente, G.; Gerritsen, J.W.; Wiegerinck, W.; Pérez Bejerano, D.; et al. The potential of a portable, point-of-care electronic nose to diagnose tuberculosis. *J. Infect.* **2017**, *75*, 441–447. [CrossRef] [PubMed]

76. Mohamed, E.I.; Mohamed, M.A.; Moustafa, M.H.; Abdel-Mageed, S.M.; Moro, A.M.; Baess, A.I.; El-Kholy, S.M. Qualitative analysis of biological tuberculosis samples by an electronic nose-based artificial neural network. *Int. J. Tuberc. Lung Dis.* **2017**, *21*, 810–817. [CrossRef] [PubMed]

Review

Use of Electronic Noses in Seawater Quality Monitoring: A Systematic Review

Alessandro Tonacci *, Francesco Sansone, Raffaele Conte and Claudio Domenici

National Research Council of Italy, Institute of Clinical Physiology (CNR-IFC), Via Moruzzi 1, 56124 Pisa, Italy; francesco.sansone@ifc.cnr.it (F.S.); raffaele.conte@ifc.cnr.it (R.C.); domenici@ifc.cnr.it (C.D.)
* Correspondence: atonacci@ifc.cnr.it; Tel.: +39-050-315-2175

Received: 29 October 2018; Accepted: 19 November 2018; Published: 23 November 2018

Abstract: Electronic nose (eNose) systems are particularly appreciated for their portability, usability, relative low cost, and real-time or near real-time response. Their application finds space in several domains, including environmental monitoring. Within this field, marine monitoring is of particular scientific relevance due to the fragility of this specific environment, daily threatened by human activities that can potentially bring to catastrophic and irreversible consequences on marine wildlife. Under such considerations, a systematic review, complying with the PRISMA guidelines, was conducted covering the period up to 15 October 2018, in PubMed, ScienceDirect, and Google Scholar. Despite the relatively low number of articles published on this specific topic and the heterogeneity of the technological approaches employed, the results obtained by the various groups highlight the positive contribution eNose has given and can provide in near future for the monitoring and safeguarding of this delicate environment.

Keywords: chemical sensors; eNose; environmental monitoring; seawater; sensors; volatile organic compounds

---

1. Introduction

Anthropogenic activities produce every day large quantities of pollutants that are discharged in the environment [1,2], including oceans, and in the seawater at large. Such discharges, together with the exploitation of the sea resources nowadays represent a significant threat for the marine environment and cause a continuous, unprecedented degradation of oceans, seas and coastal areas [3–5]. The monitoring of such compounds is a pivotal action when aiming at promoting the preservation of marine and coastal areas and the sustainability of the ecosystem, at large [6].

To this extent, the approaches commonly used are based on traditional water and air quality evaluation methods, including analytical tools that, however, commonly operate at laboratory settings, and are therefore (i) unable to provide real-time or near real-time results, and (ii) possibly being associated with the occurrence of sample degradation during transportation from the sampling site to the laboratory bench.

To properly overtake such bottlenecks, portable devices can represent a smart, mostly robust, easy-to-use, low-cost alternative, capable of bringing fast, reliable, reproducible responses, without the need for sample transportation and, likely, consequent degradation.

Among such systems, electronic nose (eNose)-like tools are emerging as a popular alternative, maximizing the above mentioned advantages and keeping, at the same time, low drawbacks.

The first evidences for eNose systems dates back to the early 1980s, when Persaud and Dodd [7] first attempted mimicking the functioning of the mammalian olfactory system by means of an "electronic nose". Since then, a plethora of systems has been realized for several different applications, including biomedical and diagnostic, environmental, as well as food industry/food quality usage [8,9].

Under such considerations, here we present a systematic review to investigate the use of eNose systems within a specific field of application, represented by the monitoring of seawater pollution. To this extent, we will first outline the search strategy adopted for this work, then we will present the results obtained, critically discussing such evidences, and highlighting pros and cons of the single approaches.

2. Materials and Methods

Search Strategy

A systematic review of the literature, covering the period up to 15 October 2018, was conducted in PubMed, ScienceDirect, and Google Scholar database, according to the PRISMA guidelines [10]. The search strategy was as follows: ("eNose" or "electronic nose") and ("seawater pollution" or "seawater monitoring" or "seawater" or "pollution"). The search was limited to research articles published in English language in peer-reviewed journals or in international conferences' proceedings. After having discarded multiple hits, the results obtained were sorted by relevance and the most significant works dealing with seawater monitoring by means of eNose systems were selected (as displayed in Figure 1). Given the relatively low number of systems specifically designed for seawater monitoring, some tools designed for water monitoring in general, but adaptable to marine monitoring, were also included and critically discussed.

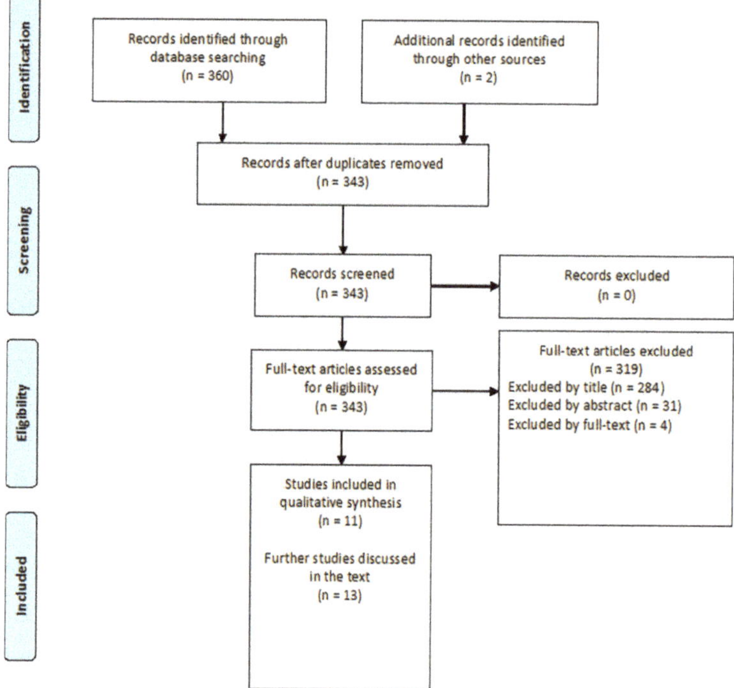

Figure 1. Flowchart for literature review.

3. Results

This systematic review retrieved a handful articles directly dealing with seawater monitoring by eNose systems. The articles taken into account are displayed in Table 1.

Table 1. Study selected (AUV: autonomous underwater vehicle; FGC: fast gas chromatography; GC-MS: gas chromatography—mass spectrometry; GSM: geosmin; IoT: Internet of Things; MIB: 2-methylisoborneol; MOS: metal oxide semiconductor; ORs: olfactory receptors; piD: photo-ionization detector(s); swCNT-FET: single-walled carbon nanotube field-effect transistor).

Study	Design	Sensor(s) Type	Monitoring Site
Bourgeois et al. [11]	(i) Sampling vessel, eNose sensor array and PC for data analysis; (ii) eNose sensor array, PC for data storage, data transfer system	(i) eNose5000; (ii) ProSAT	(i) Laboratory trials; (ii) Cranfield University Sewage Works facilities
Tzing et al. [12]	eNose used to identify source of oil leakage from an accident site. zNose, GC-MS used for results' confirmation and quantitative analysis	eNose: Cyranose 320; zNose FGC/SAW 7100; GC-MS: Varian CP-3800 + Saturn 200 ion-trap	On-field (accident site)
Goschnik et al. [13]	eNose used to discriminate between clear and polluted (chloroform and ammonia) water	Semiconductor-based KAMINA eNose	Laboratory trials
Lozano et al. [14,15]	Portable wireless resistive sensor-based eNose. electronic pump and valve, electronics and rechargeable batteries. Measurements on clear water and 11 pollutants	Home-made eNose capable of hosting resistive sensors	Laboratory trials
Tonacci et al. [16,17]	eNose, electronic pump and valve, embedded electronics and PC for data acquisition, integrated within an AUV	eNose composed of three piD sensors	On-field acquisition (La Spezia Gulf and Enfola Bay, Italy)
Son et al. [18]	bio-eNose tested to distinguish the presence of GSM and MIB	Human ORs and swCNT-FET	Laboratory trials with tap water, bottled water and river water
Moroni et al. [19]	eNose, electronic pump and valve, embedded electronics and PC for data acquisition, integrated within a moored buoy	eNose composed of three piD sensors	On-field acquisition (La Spezia Gulf and Enfola Bay, Italy)
Climent et al. [20]	eNose, architecture for data acquisition, storage. processing and user interfacing parts	eNose composed of four, IoT suitable, MOS sensors	Laboratory trials with water polluted with dimethyl disulphide, dimethyl diselenide, sulphur
Aliaño-González et al. [21]	eNose combined with chemometrics; linear discrimination analysis for classification	AlphaMOS eNose composed of MOS sensors	Laboratory trials on 444 samples from gasoline, diesel and paraffin, subjected to a natural weathering process by evaporation

Specifically focusing on the results reported in Table 1, one of the first studies performed in this field was conducted by Bourgeois and colleagues [11] in the UK. The work described the development and use of an eNose system for monitoring water and wastewater, potentially usable also for seawater assessment. The first prototype was based on an eNose5000 system consisting of an array of 12 conducting polymers, interfaced with a sampling vessel and with a PC for data analysis. Aside from this laboratory prototype, a second system was realized, with a ProSAT sensor array with eight conducting polymers as the sensing system, a PC for data storage and a data transfer system. This second system was tested on-field at the Cranfield University Sewage Works facilities and, through a proper conditioning in a temperature-controlling environment, has been demonstrated to be able to reduce humidity variations, at the same time improving reproducibility. The system was capable of detecting the presence of low concentrations (around 5 ppm) of 2-chlorophenol, with optimal performances at 25.5 °C of temperature, 170 mL/min of flow, and 0 of porosity.

As widely known, hydrocarbons are among the major pollutants of the marine environment, with oil spills occurring throughout the seas and the oceans as a consequence of marine accidents, petroleum pipeline leakages and fraudulent events. Therefore, systems capable of specifically assessing the presence of oil spills on the sea surface can be critical to reduce the environmental impact of such discharges in marine and coastal environment.

Under such consideration, Tzing and colleagues [12] developed a system, based on the eNose technology, to identify the source of spilled oil in an accident site, comparing it with a "gold standard" system composed by a gas chromatography—mass spectrometry (GC-MS) device that served to ensure the correct classification of the source site.

The system developed was composed of a Cyranose 320 eNose system, composed of an array of 32 polymer sensors, a pump for air inlet and purge, as well as a PC equipped with a dedicated software for data acquisition and analysis. Together with this system, a so-called z-Nose system was employed, constituted by an Electronic Sensor Technology FGC/SAW 7100 eNose, equipped with a pre-concentrating trap containing Tenax absorbent, a short gas chromatography (GC) column, a pneumatic control, a fast SAW detector, as well as a programmable gate array microprocessor. As for reference analysis, a Varian CP-3800 GC connected to a Saturn 2000 ion-trap mass spectrometer (MS) was used.

Spilled oil samples were collected from surface water at the site of an accident, whereas a set of known compounds, including gasoline, jet fuel, diesel, and fuel oil were also collected, properly stored and transported to the laboratory, representing the reference samples for the eNose (1 mL of oil loaded into a 30 mL vial, equilibrated at ambient condition for 5 min).

By means of a PCA, the eNose classified the unknown stimulus as "jet fuel", as further agreed by zNose and GC-MS systems, demonstrating the good classification capability of the eNose system. However, the main limitation of this promising approach resides in the fact that the test set employed was represented by only one sample that, despite the high internal consistency of the eNose response, could lead to misleading conclusions.

A couple of years later, Goschnik et al. [13] employed the semiconductor-based eNose system KAMINA to analyze water samples polluted with chloroform (chloro-organic solvents-like) or ammonia (fecal contamination-like). Despite this specific trial was mainly dealing with wastewaters, this application could have had the potentialities to be translated into a system usable for seawater monitoring, therefore, it was added to the present investigation. The trials were performed with a flow rate of 500 mL/min and an entrance temperature of 20 °C, subsequently increased up to 240 °C and 300 °C at the two microarray ends, respectively. The results obtained for this investigation, showing a detection limit below 1 ppm, demonstrated the usefulness of the KAMINA system for this purpose, however revealing a strong need for building up a populated, reliable reference database to support the correct classification by the eNose system. The absence of such a support would dramatically decrease the performances of the system, as frequently occurs with eNose approaches.

A very interesting approach was proposed by Lozano et al. in 2014 [14] and updated a few years later [15]. There, the authors proposed a portable eNose system equipped with an IEEE 802.11 transceiver for wireless communication in order to be ready for inclusion within a wide eNose network for distributed measurements. The system developed, also featuring an electronic pump and valve, as well as embedded electronics and rechargeable batteries, was able to work with several different resistive microsensors. By performing measurements on cycles of 60 s of adsorption and 540 s of desorption by means of an headspace sampling system, the prototype presented was demonstrated to be able to reliably discriminate between pollutants (especially when a relatively low number of compounds are taken into account) using principal component analysis (PCA) and a few different artificial neural networks (ANNs). More specifically, the compounds used to evaluate the response of the system included: blank water, acetone, toluene, ammonia, formaldehyde, hydrogen peroxide, ethanol, benzene, dichloromethane, acetic acid, xylene and dimethylacetamide. The correct discrimination occurred in more than 90% of samples in most cases, with concentrations of the various compounds of around 1 mL in a 20 mL vial at 16 °C constant temperature.

Specifically dealing with seawater pollution, the work published by Tonacci and colleagues [16] described the implementation of an eNose system to be used within an autonomous underwater vehicle (AUV) for monitoring oil leakages in a marine protected area, including the Tuscan Archipelago and Cetacean Sanctuary, North Mediterranean. The system implemented was composed by a sensing part, featuring Photo-Ionization Detectors (piD), a system for air inlet and purge made up of an electronic pump and a valve, and a smart embedded electronics, based on an Arduino Mega 2560 board. A PC for data acquisition was also foreseen, and ANNs of the type Kohonen self-organizing map (KSOM) were implemented and trained in order to: (i) recognize the level of pollution, independently from its source (e.g., high, medium, low level), and (ii) identify the substance detected within a set of known compounds (e.g., crude oil, diesel fuel, gasoline, jet fuel). Despite the good accuracy of classifiers, especially for the first KSOM (around 74% of correct classification for the first KSOM, 67% for the second one, and a fast response (below 20 s), with pollutants' concentration of 10^{-3} vol/vol), the system was seen to provide not reliable response when exposed to relative humidity (RH) above 70%, thus representing a noteworthy limitation of the tool described. Two further expansions of this application were also presented by the same group. The first one adopted the same approach including the eNose system within a moored buoy in order to combine dynamic (AUV) and static (buoy) monitoring of a given marine area [19], whereas the second one scaled the payload system in order to make room for future inclusion of such system within a real-time context of distributed monitoring [17].

A novel approach to the problem was adopted by Son and colleagues in 2015 [18]. They developed a bio-eNose consisting of human olfactory receptors, subcloned in pcDNA3 mammalian expression vectors containing the first 20 amino acids of human rhodopsin (Rho-tag), and single-walled carbon nanotube field-effect transistor (swCNT-FET). The system functioning was assessed to distinguish the presence of geosmin and 2-methylisoborneol, compounds produced by bacteria and reliable indicators of contamination in the water supply system. The approach adopted can be tailored upon the users' needs also to investigate the presence of contaminants in seawater. The system was demonstrated to be reliable in detecting the proposed stimuli at concentration as low as 10 ng/L, with the possibility to assess the presence of both compounds at the same time, thanks to the specific binding between hOR51S1 and hOR3A4 and geosmin and 2-methylisoborneol, respectively. Very importantly, the correct classification of the eNose system does not require a specific pre-treatment of water samples, allowing the use of this solution for rapid monitoring of water quality. The binding between other olfactory receptors and specific pollutants could enable applying the solution to other, customized applications and experimental settings.

Recent technological advances allowed researchers to explore new solutions in this specific field. Internet of Things (IoT), for example, represents a landmark revolution in the field of sensing and bio-sensing. This promising and extremely actual approach was employed by Climent and colleagues [20], which developed a low-cost, portable eNose, named Multisensory Odor Olfactory System (MOOSY4), for water quality assessment purposes. In that work, the MOOSY4 was employed for the quality control of bottled water, but can be also used for seawater monitoring. The prototype was composed of four metal oxide semiconductor (MOS) gas sensors, suitable for the IoT technology, with a system architecture featuring data acquisition, storage, processing and user interfacing parts. The volatiles monitored, dimethyl disulphide, dimethyl diselenide, sulphur, were correctly recognized in 86% of cases, making the system extremely suitable for the purpose it was conceived for, and potentially applicable to a wide range of volatiles monitoring.

As hydrocarbons are, as reported, the main source of marine environmental pollution, specific approaches have been recently adopted tailored at this peculiar domain. For example, Aliaño-González and colleagues [21] used an AlphaMOS eNose combined with chemometrics to identify and discriminate 444 samples composed 40 µL of different petroleum-derived products (PDPs), including gasoline, diesel and paraffin, poured on a support and subjected to a natural weathering process by evaporation for one month. Taking advantage of the use of the linear discriminant analysis allowed the scientists to perform a correct discrimination of presence/absence of PDP in all cases and of the different PDPs in 97.7% of samples.

4. Discussion

The problem of environmental pollution is central in today's society, and full attention is paid by both National and International Environmental Organisms, as well as researchers, in finding solutions that can properly respond to the pressing necessity of environmental safeguarding. The extremely fragile ecosystem represented by marine and coastal areas is continuously threatened by manmade activities (e.g., ship transits, industrial discharges, etc.) that, in the medium- to long-term, can result in serious and irreversible consequences for marine areas and their wildlife [22,23].

To this extent, several initiatives, including national and international research projects, task-forces, position papers, and regulations have been promoted in last decades, many of which highlighting the need for tailored interventions both in a more preventive sense (monitoring of oil spills, etc.) and concerning accident remediation [24]. On the edge between these two phases, a number of smart monitoring systems have been developed in many research centers, to ensure a proper monitoring of a given area and, at the same time, allowing the possibility to trigger eventual alarms when the concentration of pollutants or, more specifically, of a single pollutant, exceeds a given safety threshold. Such solutions are especially useful in marine protected areas, where even the detection of very low levels of pollutants could enable the competent authorities to undergo quick, effective countermeasures, often preventing real environmental disasters. Within this framework, the use of low-cost, sensitive tools, ensuring a fast, real-time, or near real-time response is of particular interest, and this fact paved the way for the employment of eNose systems for this purpose.

In the last years, several prototypes and products based on this technology have been developed and employed for seawater monitoring, using different technologies as time flowed.

Among the most widely used technologies, especially in the first works published, are conducting polymers [11,12], providing good results, but in some cases obtained on a small number of trials or compounds. However, it is worth noting that such sensors are largely affected by environmental conditions, that could dramatically impact on the recognition performances of the tool, overall. The experimental setting is also pivotal with this approach, since the sensors should often be carefully cleaned after each measurement, reducing their applicability in a real environmental setting for continuous measurements unless a coupled purging device is used.

Semiconductors, which are the basis of most eNose systems developed to date, were also employed several times to this extent [13,20,21] and, considering the timeline of existing publications on the argument, are probably the most actual approach to the problem. Some contrasting results are present under this technology (e.g., considering the limits highlighted by the work by Goschnik and colleagues), however, the use of such devices probably guarantees the best performances, to date. The main limitation identified under this approach is the need for a well populated reference database to train the related recognition algorithms, but with the technological developments of recent times the issue was, in most cases, successfully resolved.

Other approaches retrieved in this domain included the use of photo-ionization detectors [16,17,19], featuring very fast response times, however displaying significant issues with high humidity environments and being characterized by high costs, and resistive detectors like the ones employed by Lozano and colleagues [14] and Herrero and colleagues [15]. The last solution mentioned is characterized by the strong need to control environmental conditions for the signal stability, a common problem with most of the other technologies normally employed in this specific field.

Additionally, promising hybrid approaches including a biological part, consisting of human olfactory receptors, as well as a technological part, were successfully employed by Son and colleagues [18], possibly opening the way for new explorations to reduce the impact of one of the most known issues of the eNose systems, that is non-specificity. In fact, thanks to the selective bond between compound and receptor, this approach could enable the detection of specific compounds, even though the reversibility of this bond remains questionable. The provision of such devices for new, promising technologies, including the IoT, as already experimented by Climent et al. [20], could enable new perspectives, at large in environmental, and specifically in seawater, monitoring.

Author Contributions: A.T. defined the rationale of the work, performed the literature search, critically revised the selected articles, and wrote the manuscript; F.S. participated in the bias and quality assessment, discussed the results obtained, and critically revised the manuscript; R.C. participated in the bias and quality assessment, discussed the results obtained, and critically revised the manuscript; and C.D. participated in the definition of the work rationale, discussed the results obtained and critically revised the manuscript. All the authors approved the final version of the manuscript.

Funding: This research received no external funding.

Conflicts of Interest: The authors declare no conflict of interest.

References

1. Correa, S.M.; Arbilla, G.; Marques, M.R.C.; Oliveira, K.M.P.G. The impact of btex emissions from gas stations into the atmosphere. *Atmos. Pollut. Res.* **2012**, *3*, 163–169. [CrossRef]

2. Lapworth, D.J.; Baran, N.; Stuart, M.E.; Ward, R.S. Emerging organic contaminants in groundwater: A review of sources, fate and occurrence. *Environ. Pollut.* **2012**, *163*, 287–303. [CrossRef] [PubMed]

3. Bagby, S.C.; Reddy, C.M.; Aeppli, C.; Fisher, G.B.; Valentine, D.L. Persistence and biodegradation of oil at the ocean floor following deepwater horizon. *Proc. Natl. Acad. Sci. USA* **2017**, *114*, E9–E18. [CrossRef] [PubMed]

4. Chocarro-Ruiz, B.; Herranz, S.; Fernández Gavela, A.; Sanchís, J.; Farré, M.; Marco, M.P.; Lechuga, L.M. Interferometric nanoimmunosensor for label-free and real-time monitoring of Irgarol 1051 in seawater. *Biosens. Bioelectron.* **2018**, *117*, 47–52. [CrossRef] [PubMed]

5. Evans, M.; Liu, J.; Bacosa, H.; Rosenheim, B.E.; Liu, Z. Petroleum hydrocarbon persistence following the deepwater horizon oil spill as a function of shoreline energy. *Mar. Pollut. Bull.* **2017**, *115*, 47–56. [CrossRef] [PubMed]

6. Bakker, K. Water Security: Research Challenges and Opportunities. *Science* **2012**, *337*, 914–915. [CrossRef] [PubMed]

7. Persaud, K.; Dodd, G. Analysis of discrimination mechanisms in the mammalian olfactory system using a model nose. *Nature* **1982**, *299*, 352–355. [CrossRef] [PubMed]

8. Tonacci, A.; Sansone, F.; Pala, A.P.; Conte, R. Exhaled breath analysis in evaluation of psychological stress: A short literature review. *Int. J. Psychol.* **2018**. [CrossRef] [PubMed]

9. Wojnowski, W.; Majchrzak, T.; Dymerski, T.; Gębicki, J.; Namieśnik, J. Portable Electronic Nose Based on Electrochemical Sensors for Food Quality Assessment. *Sensors* **2017**, *17*, 2715. [CrossRef] [PubMed]

10. Moher, D.; Liberati, A.; Tetzlaff, J.; Altman, D.G.; The PRISMA Group. Preferred reporting items for systematic reviews and meta-analyses: The PRISMA statement. *PLoS Med.* **2009**, *6*, e1000097. [CrossRef] [PubMed]

11. Bourgeois, W.; Hogben, P.; Pike, A.; Stuetz, R.M. Development of a sensor array based measurement system for continuous monitoring of water and wastewater. *Sens. Actuators B* **2003**, *88*, 312–319. [CrossRef]
12. Tzing, S.H.; Chang, J.Y.; Ghule, A.; Chang, J.J.; Lo, B.; Ling, Y.C. A simple and rapid method for identifying the source of spilled oil using an electronic nose: Confirmation by gas chromatography with mass spectrometry. *Rapid Commun. Mass Spectrom.* **2003**, *17*, 1873–1880. [CrossRef] [PubMed]
13. Goschnick, J.; Koronczi, I.; Frietsch, M.; Kiselev, I. Water pollution recognition with the electronic nose KAMINA. *Sens. Actuators B* **2005**, *106*, 182–186. [CrossRef]
14. Lozano, J.; Santos, J.P.; Suárez, J.I.; Arroyo, P.; Herrero, J.L.; Martín, A. Detection of pollutants in water samples with a wireless hand-held e-nose. *Procedia Eng.* **2014**, *87*, 556–559. [CrossRef]
15. Herrero, J.L.; Lozano, J.; Santos, J.P.; Suárez, J.I. On-line classification of pollutants in water using wireless portable electronic noses. *Chemosphere* **2016**, *152*, 107–116. [CrossRef] [PubMed]
16. Tonacci, A.; Corda, D.; Tartarisco, G.; Pioggia, G.; Domenici, C. A smart sensor system for detecting hydrocarbon Volatile Organic Compounds in sea water. *CLEAN Soil Air Water* **2015**, *43*, 147–152. [CrossRef]
17. Tonacci, A.; Lacava, G.; Lippa, M.A.; Lupi, L.; Cocco, M.; Domenici, C. Electronic Nose and AUV: A Novel Perspective in Marine Pollution Monitoring. *Mar. Technol. Soc. J.* **2015**, *49*, 18–24. [CrossRef]
18. Son, M.; Cho, D.G.; Lim, J.H.; Park, J.; Hong, S.; Ko, H.J.; Park, T.H. Real-time monitoring of geosmin and 2-methylisoborneol, representative odor compounds in water pollution using bioelectronics nose with human-like performance. *Biosens. Bioelectron.* **2015**, *74*, 199–206. [CrossRef] [PubMed]
19. Moroni, D.; Pieri, G.; Salvetti, O.; Tampucci, M.; Domenici, C.; Tonacci, A. Sensorized buoy for oil spill early detection. *Methods Oceanogr.* **2016**, *17*, 221–231. [CrossRef]
20. Climent, E.; Pelegri-Sebastia, J.; Sogorb, T.; Talens, J.B.; Chilo, J. Development of the MOOSY4 eNose IoT for Sulphur-Based VOC Water Pollution Detection. *Sensors* **2017**, *17*, 1917. [CrossRef] [PubMed]
21. Aliaño-González, M.J.; Ferreiro-González, M.; Barbero, G.F.; Ayuso, J.; Álvarez, J.A.; Palma, M.; Barroso, C.G. An Electronic Nose Based Method for the Discrimination of Weathered Petroleum-Derived Products. *Sensors* **2018**, *18*, 2180. [CrossRef] [PubMed]
22. Mille, G.; Asia, L.; Guiliano, M.; Malleret, L.; Doumenq, P. Hydrocarbons in coastal sediments from the Mediterranean sea (Gulf of Fos area, France). *Mar. Pollut. Bull.* **2007**, *54*, 566–575. [CrossRef] [PubMed]
23. Zrafi, I.; Bakhrouf, A.; Mahmoud, R.; Dalila, S.D. Aliphatic and aromatic biomarkers for petroleum hydrocarbon monitoring in Khniss Tunisian-Coast, (Mediterranean Sea). *Procedia Environ. Sci.* **2013**, *18*, 211–220. [CrossRef]
24. Azzellino, A.; Panigada, S.; Lanfredi, C.; Zanardelli, M.; Airoldi, S.; Notarbartolo di Sciara, G. Predictive habitat models for managing marine areas: Spatial and temporal distribution of marine mammals within the Pelagos Sanctuary (Northwestern Mediterranean sea). *Ocean Coast. Manag.* **2012**, *67*, 63–74. [CrossRef]

MDPI

St. Alban-Anlage 66

4052 Basel

Switzerland

Tel. +41 61 683 77 34

Fax +41 61 302 89 18

www.mdpi.com

Biosensors Editorial Office

E-mail: biosensors@mdpi.com

www.mdpi.com/journal/biosensors

www.ingramcontent.com/pod-product-compliance
Lightning Source LLC
LaVergne TN
LVHW070632170726
843515LV00004B/232